Mathematical
Olympiad
in China (2017–2018)

Problems and Solutions

Mathematical Olympiad Series

ISSN: 1793-8570

Series Editors: Lee Peng Yee *(Nanyang Technological University, Singapore)*
Xiong Bin *(East China Normal University, China)*

Published

Vol. 18 *Mathematical Olympiad in China (2017–2018):*
Problems and Solutions
edited by Bin Xiong (East China Normal University, China)

Vol. 17 *Mathematical Olympiad in China (2015–2016):*
Problems and Solutions
edited by Bin Xiong (East China Normal University, China)

Vol. 16 *Sequences and Mathematical Induction:*
In Mathematical Olympiad and Competitions
Second Edition
by Zhigang Feng (Shanghai Senior High School, China)
translated by: Feng Ma, Youren Wang

Vol. 15 *Mathematical Olympiad in China (2011–2014):*
Problems and Solutions
edited by Bin Xiong (East China Normal University, China) &
Peng Yee Lee (Nanyang Technological University, Singapore)

Vol. 14 *Probability and Expectation*
by Zun Shan (Nanjing Normal University, China)
translated by: Shanping Wang (East China Normal University, China)

Vol. 13 *Combinatorial Extremization*
by Yuefeng Feng (Shenzhen Senior High School, China)

Vol. 12 *Geometric Inequalities*
by Gangsong Leng (Shanghai University, China)
translated by: Yongming Liu (East China Normal University, China)

Vol. 11 *Methods and Techniques for Proving Inequalities*
by Yong Su (Stanford University, USA) &
Bin Xiong (East China Normal University, China)

The complete list of the published volumes in the series can be found at
http://www.worldscientific.com/series/mos

Vol. 18 | Mathematical Olympiad Series

Mathematical Olympiad

in China (2017–2018)

Problems and Solutions

Editor-in-Chief
Xiong Bin
East China Normal University, China

English Translators
Wang Shanping
Journal of East China Normal University, China
Chen Haoran
Xi'an Jiaotong-Liverpool University

Copy Editors
Ni Ming
Kong Lingzhi
East China Normal University Press, China

East China Normal University Press

World Scientific

Published by

East China Normal University Press
3663 North Zhongshan Road
Shanghai 200062
China

and

World Scientific Publishing Co. Pte. Ltd.
5 Toh Tuck Link, Singapore 596224
USA office: 27 Warren Street, Suite 401-402, Hackensack, NJ 07601
UK office: 57 Shelton Street, Covent Garden, London WC2H 9HE

Library of Congress Cataloging-in-Publication Data
Names: Xiong, Bin, editor.
Title: Mathematical Olympiad in China (2017–2018) : problems and solutions /
     editor Xiong Bin, East China Normal University, China.
Description: Shanghai, China : East China Normal University Press ; Hackensack, NJ :
     World Scientific, [2023] | Series: Mathematical olympiad series, 1793-8570 ; vol. 18
Identifiers: LCCN 2022013726 | ISBN 9789811256295 (hardcover) |
     ISBN 9789811257384 (paperback) | ISBN 9789811256301 (ebook) |
     ISBN 9789811256318 (ebook other)
Subjects: LCSH: International Mathematical Olympiad. | Mathematics--Problems, exercises, etc. |
     Mathematics--Competitions--China.
Classification: LCC QA43 .M31454 2023 | DDC 510.76--dc23/eng/20220330
LC record available at https://lccn.loc.gov/2022013726

British Library Cataloguing-in-Publication Data
A catalogue record for this book is available from the British Library.

For any available supplementary material, please visit
https://www.worldscientific.com/worldscibooks/10.1142/12836#t=suppl

Typeset by Stallion Press
Email: enquiries@stallionpress.com

Printed in Singapore

Preface

The first time China participated in IMO was in 1985, when two students were sent to the 26th IMO. Since 1986, China has a team of 6 students at every IMO except in 1998 when it was held in Taiwan. So far, up to 2018, China has achieved the number one ranking in team effort 24 times. A great majority of students received gold medals. The fact that China obtained such encouraging results is due to, on one hand, Chinese students' hard work and perseverance, and on the other hand, the effort of the teachers in schools and the training offered by national coaches. We believe this is also a result of the education system in China, in particular, the emphasis on training of the basic skills in science education.

The materials of this book come from two volumes (Vol. 2017 and Vol. 2018) of a book series in Chinese "走向 IMO: 数学奥林匹克试题集锦" (*Forward to IMO: A Collection of Mathematical Olympiad Problems*). It is a collection of problems and solutions of the major mathematical competitions in China. It provides a glimpse of how the China national team is selected and formed. First, there is the China Mathematical Competition, a national event. It is held on the second Sunday of September every year. Through the competition, about 380 students are selected to join the China Mathematical Olympiad (commonly known as the winter camp), or in short CMO, in November. CMO lasts for five days. Both the type and the difficulty of the problems match those of IMO. Similarly, students are given three problems to solve in 4.5 hours each day. From CMO, 60 students are selected to form a national training team. The training takes place for two weeks in the month of March. After four to six tests, plus two qualifying examinations, six students are finally selected to form the national team, taking part in IMO in July of that year.

In view of the differences in education, culture and economy of the western part of China with the coastal part in eastern China, mathematical competitions in West China did not develop as fast as the rest of the country. In order to promote the activity of mathematical competition, and to enhance the level of mathematical competition, starting from 2001, China Mathematical Olympiad Committee organizes the China Western Mathematical Olympiad.

Since 2012, the China Western Mathematical Olympiad has been renamed the China Western Mathematical Invitation. The competition dates have been changed from the first half of October to the middle of August since 2013.

The development of this competition reignited the enthusiasm of Western students for mathematics. Once again, the figures of Western students often appeared in the national team.

Since 1995, there was no female student in the Chinese national team. In order to encourage more female students participating in the mathematical competition, starting from 2002, China Mathematical Olympiad Committee has been conducting the China Girls' Mathematical Olympiad. Again, the top twelve winners will be admitted directly into the CMO.

The authors of this book are coaches of the China national team. They are Xiong Bin, Wu Jianping, Leng Gangsong, Yu Hongbing, Yao Yijun, Qu Zhenhua, Li Ting, Ai Yinhua, Wang Bin, Fu Yunhao, et al., He Yijie and Zhang Sihui. Those who took part in the translation work are Chen Haoran and Wang Shanping. We are grateful to Qiu Zonghu, Wang Jie, Zhou Qin, Wu Jianping, and Pan Chengbiao for their guidance and assistance to the authors. We are grateful to Ni Ming and Kong Linzhi of East China Normal University Press. Their effort has helped make our job easier. We are also grateful to Tan Rok Ting and Liu Nijia of World Scientific Publishing for their hard work leading to the final publication of the book.

Authors
October 2021

Introduction

Early days

The International Mathematical Olympiad (IMO), founded in 1959, is one of the most competitive and highly intellectual activities in the world for high school students.

Even before IMO, there were already many countries which had mathematics competition. They were mainly the countries in Eastern Europe and in Asia. In addition to the popularization of mathematics and the convergence in educational systems among different countries, the success of mathematical competitions at the national level provided a foundation for the setting-up of IMO. The countries that asserted great influence are Hungary, the former Soviet Union, and the United States. Here is a brief history of the IMO and mathematical competition in China.

In 1894, the Department of Education in Hungary passed a motion and decided to conduct a mathematical competition for the secondary schools. The well-known scientist, *J. von Etövös*, was the Minister of Education at that time. His support in the event had made it a success and thus it was well publicized. In addition, the success of his son, *R. von Etövös*, who was also a physicist, in proving the principle of equivalence of the general theory of relativity by *A. Einstein* through experiment, had brought Hungary to the world stage in science. Thereafter, the prize for mathematics competition in Hungary was named *"Etövös* prize". This was the first formally organized mathematical competition in the world. In what follows, Hungary had indeed produced a lot of well-known scientists including *L. Fejér*, *G. Szegö*, *T. Radó*, *A. Haar* and *M. Riesz* (in real analysis), *D. König* (in combinatorics), *T. von Kármán* (in aerodynamics), and *J. C. Harsanyi* (in game theory), who had also won the Nobel Prize for Economics in 1994. They all were the winners of Hungary mathematical competition.

The top scientific genius of Hungary, *J. von Neumann*, was one of the leading mathematicians in the 20th century. *Neumann* was overseas while the competition took place. Later he did the competition himself and it took him half an hour to complete. Another mathematician worth mentioning is the highly productive number theorist *P. Erdös*. He was a pupil of *Fejér* and a winner of the Wolf Prize. *Erdös* was very passionate about mathematical competition and setting competition questions. His contribution to discrete mathematics was unique and greatly significant. The rapid progress and development of discrete mathematics over the subsequent decades had indirectly influenced the types of questions set in IMO. An internationally recognized prize was named after *Erdös* to honour those who had contributed to the education of mathematical competition. Professor *Qiu Zonghu* from China had won the prize in 1993.

In 1934, a famous mathematician *B. Delone* conducted a mathematical competition for high school students in Leningrad (now St. Petersburg). In 1935, Moscow also started organizing such events. Other than being interrupted during the World War II, these events had been carried on until today. As for the Russian Mathematical Competition (later renamed as the Soviet Mathematical Competition), it was not started until 1961. Thus, the former Soviet Union and Russia became the leading powers of Mathematical Olympiad. A lot of grandmasters in mathematics including the great *A. N. Kolmogorov* were all very enthusiastic about the mathematical competition. They would personally involve themselves in setting the questions for the competition. The former Soviet Union even called it the Mathematical Olympiad, believing that mathematics is the "gymnastics of thinking". These points of view gave a great impact on the educational community. The winner of the Fields Medal in 1998, *M. Kontsevich*, was once the first runner-up of the Russian Mathematical Competition. *G. Kasparov*, the international chess grandmaster, was once the second runner-up. *Grigori Perelman*, the winner of the Fields Medal in 2006 (but he declined), who solved the Poincaré's Conjecture, was a gold medalist of IMO in 1982.

In the United States of America, due to the active promotion by the renowned mathematician *G. D. Birkhoff* and his son, together with *G. Pólya*, the Putnam mathematics competition was organized in 1938 for junior undergraduates. Many of the questions were within the scope of high school students. The top five contestants of the Putnam mathematical competition would be entitled to the membership of Putnam. Many of these were eventually outstanding mathematicians. There were the famous *R. Feynman* (winner of the Nobel Prize for Physics, 1965), *K. Wilson* (winner of the Nobel Prize for Physics, 1982), *J. Milnor* (winner of the Fields Medal, 1962), *D. Mumford* (winner of the Fields Medal, 1974), and *D. Quillen* (winner of the Fields Medal, 1978).

Since 1972, in order to prepare for the IMO, the United States of America Mathematical Olympiad (USAMO) was organized. The standard of questions posed was very high, parallel to that of the Winter Camp in China. Prior to this, the United States had organized American High School Mathematics Examination (AHSME) for the high school students since 1950. This was at the junior level and yet the most popular mathematics competition in America. Originally, it was planned to select about 100 contestants from AHSME to participate in USAMO. However, due to the discrepancy in the level of difficulty between the two competitions and other restrictions, from 1983 onwards, an intermediate level of competition, namely, American Invitational Mathematics Examination (AIME), was introduced. Henceforth both AHSME and AIME became internationally well-known. Since 2000, AHSME was replaced by AMC 12 and AMC 10. Students who perform well on the AMC 12 and AMC 10 are invited to participate in AIME. The combined scores of the AMC 12 and the AIME are used to determine approximately 270 individuals that will be invited back to take the USAMO, while the combined scores of the AMC 10 and the AIME are used to determine approximately 230 individuals that will be invited to take the USAJMO (United States of America Junior Mathematical Olympiad), which started in 2010 and follows the same format as the USAMO. A few cities in China had participated in the competition and the results were encouraging.

Similarly, as in the former Soviet Union, the Mathematical Olympiad education was widely recognized in America. The book "How to Solve it" written by *George Polya* along with many other titles had been translated into many different languages. *George Polya* provided a whole series of general heuristics for solving problems of all kinds. His influence in the educational community in China should not be underestimated.

International Mathematical Olympiad

In 1956, the East European countries and the Soviet Union took the initiative to organize the IMO formally. The first International Mathematical Olympiad (IMO) was held in Brasov, Romania, in 1959. At that time, there were only seven participating countries, namely, Romania, Bulgaria, Poland, Hungary, Czechoslovakia, East Germany, and the Soviet Union.

Subsequently, the United States of America, United Kingdom, France, Germany, and also other countries including those from Asia joined. Today, the IMO had managed to reach almost all the developed and developing countries. Except in the year 1980 due to financial difficulties faced by the host country, Mongolia, there were already 59 Olympiads held and 107 countries and regions participating.

The mathematical topics in the IMO include algebra, combinatorics, geometry, number theory. These areas have provided guidance for setting questions for the competitions. Other than the first few Olympiads, each IMO is normally held in mid-July every year and the test paper consists of 6 questions in all. The actual competition lasts for 2 days for a total of 9 hours where participants are required to complete 3 questions each day. Each question is 7 points which total up to 42 points. The full score for a team is 252 marks. About half of the participants will be awarded a medal, where 1/12 will be awarded a gold medal. The numbers of gold, silver and bronze medals awarded are in the ratio of 1:2:3 approximately. In the case when a participant provides a better solution than the official answer, a special award is given.

Each participating country and region will take turn to host the IMO. The cost is borne by the host country. China had successfully hosted the 31st IMO in Beijing. The event had made a great impact on the mathematical community in China. According to the rules and regulations of the IMO, all participating countries are required to send a delegation consisting of a leader, a deputy leader and 6 contestants. The problems are contributed by the participating countries and are later selected carefully by the host country for submission to the international jury set up by the host country. Eventually, only 6 problems will be accepted for use in the competition. The host country does not provide any question. The short-listed problems are subsequently translated, if necessary, in English, French, German, Spain, Russian, and other working languages. After that, the team leaders will translate the problems into their own languages.

The answer scripts of each participating team will be marked by the team leader and the deputy leader. The team leader will later present the scripts of their contestants to the coordinators for assessment. If there is any dispute, the matter will be settled by the jury. The jury is formed by the various team leaders and an appointed chairman by the host country. The jury is responsible for deciding the final 6 problems for the competition. Their duties also include finalizing the grading standard, ensuring the accuracy of the translation of the problems, standardizing replies to written queries raised by participants during the competition, synchronizing differences in grading between the team leaders and the coordinators and also deciding on the cut-off points for the medals depending on the contestants' results as the difficulties of problems each year are different.

China had participated informally in the 26th IMO in 1985. Only two students were sent. Starting from 1986, except in 1998 when the IMO was held in Taiwan, China had always sent 6 official contestants to the IMO. Today, the Chinese contestants not only performed outstandingly in the IMO, but also in the International Physics, Chemistry, Informatics, and Biology Olympiads. This can be regarded as an indication that China pays great attention to the training of basic skills in mathematics and science education.

Winners of the IMO

Among all the IMO medalists, there were many of them who eventually became great mathematicians. They were also awarded the Fields Medal, Wolf Prize and Nevanlinna Prize (a prominent mathematics prize for computing and informatics). In what follows, we name some of the winners.

G. Margulis, a silver medalist of IMO in 1959, was awarded the Fields Medal in 1978. *L. Lovasz*, who won the Wolf Prize in 1999, was awarded the Special Award in IMO consecutively in 1965 and 1966. *V. Drinfeld*, a gold medalist of IMO in 1969, was awarded the Fields Medal in 1990. *J.-C. Yoccoz* and *T. Gowers*, who were both awarded the Fields Medal in 1998, were gold medalists in IMO in 1974 and 1981 respectively. A silver medalist of IMO in 1985, *L. Lafforgue*, won the Fields Medal in 2002. A gold medalist of IMO in 1982, *Grigori Perelman* from Russia, was awarded the Fields Medal in 2006 for solving the final step of the Poincaré conjecture. In 1986, 1987, and 1988, *Terence Tao* won a bronze, silver, and gold medal respectively. He was the youngest participant to date in the IMO, first competing at the age of ten. He was also awarded the Fields Medal in 2006. Gold medalist of IMO 1988 and 1989, *Ngo Bau Chao*, won the Fields Medal in 2010, together with the bronze medalist of IMO 1988, *E. Lindenstrauss*. Gold medalist of IMO 1994 and 1995 *Maryam Mirzakhani* won the Fields Medal in 2014. A gold medalist of IMO in 1995, Artur Avila won the Fields Medal in 2014. Gold medalist of IMO 2005, 2006 and 2007, Peter Scholze won the Fields Medal in 2018. A Bronze medalist of IMO in 1994, Akshay Venkatesh won the Fields Medal in 2018.

A silver medalist of IMO in 1977, *P. Shor*, was awarded the Nevanlinna Prize. A gold medalist of IMO in 1979, *A. Razborov*, was awarded the

Nevanlinna Prize. Another gold medalist of IMO in 1986, *S. Smirnov*, was awarded the Clay Research Award. *V. Lafforgue*, a gold medalist of IMO in 1990, was awarded the European Mathematical Society prize. He is *L. Lafforgue*'s younger brother.

Also, a famous mathematician in number theory, *N. Elkies*, who is also a professor at Harvard University, was awarded a gold medal of IMO in 1982. Other winners include *P. Kronheimer* awarded a silver medal in 1981 and *R. Taylor* a contestant of IMO in 1980.

Mathematical competition in China

Due to various reasons, mathematical competition in China started relatively late but is progressing vigorously.

"We are going to have our own mathematical competition too!" said *Hua Luogeng*. *Hua* is a household name in China. The first mathematical competition was held concurrently in Beijing, Tianjin, Shanghai, and Wuhan in 1956. Due to the political situation at the time, this event was interrupted a few times. Until 1962, when the political environment started to improve, Beijing and other cities started organizing the competition though not regularly. In the era of Cultural Revolution, the whole educational system in China was in chaos. The mathematical competition came to a complete halt. In contrast, the mathematical competition in the former Soviet Union was still on-going during the war and at a time under the difficult political situation. The competitions in Moscow were interrupted only 3 times between 1942 and 1944. It was indeed commendable.

In 1978, it was the spring of science. *Hua Luogeng* conducted the Middle School Mathematical Competition for 8 provinces in China. The mathematical competition in China was then making a fresh start and embarked on a road of rapid development. *Hua* passed away in 1985. In commemorating him, a competition named *Hua Luogeng* Gold Cup was set up in 1986 for students in Grade 6 and 7 and it has a great impact.

The mathematical competitions in China before 1980 can be considered as the initial period. The problems set were within the scope of middle school textbooks. After 1980, the competitions were gradually moving towards the senior middle school level. In 1981, the Chinese Mathematical Society decided to conduct the China Mathematical Competition, a national event for high schools.

In 1981, the United States of America, the host country of IMO, issued an invitation to China to participate in the event. Only in 1985, China sent two contestants to participate informally in the IMO. The results were not encouraging. In view of this, another activity called the Winter Camp was conducted after the China Mathematical Competition. The Winter Camp was later renamed as the China Mathematical Olympiad or CMO. The winning team would be awarded the *Chern Shiing-Shen* Cup. Based on the outcome at the Winter Camp, a selection would be made to form the 6-member national team for IMO. From 1986 onwards, other than the year when IMO was organized in Taiwan, China had been sending a 6-member team to IMO. Up to 2018, China had been awarded the overall team champion for 19 times.

In 1990, China had successfully hosted the 31st IMO. It showed that the standard of mathematical competition in China has leveled that of other leading countries. First, the fact that China achieves the highest marks at the 31st IMO for the team is evidence of the effectiveness of the pyramid approach in selecting the contestants in China. Secondly, the Chinese mathematicians had simplified and modified over 100 problems and submitted them to the team leaders of the 35 countries for their perusal. Eventually, 28 problems were recommended. At the end, 5 problems were chosen (IMO requires 6 problems). This is another evidence to show that China has achieved the highest quality in setting problems. Thirdly, the answer scripts of the participants were marked by the various team leaders and assessed by the coordinators who were nominated by the host countries. China had formed a group 50 mathematicians to serve as coordinators who would ensure the high accuracy and fairness in marking. The marking process was completed half a day earlier than it was scheduled. Fourthly, that was the first ever IMO organized in Asia. The outstanding performance by China had encouraged the other developing countries, especially those in Asia. The organizing and coordinating work of the IMO by the host country was also reasonably good.

In China, the outstanding performance in mathematical competition is a result of many contributions from all quarters of mathematical community. There are the older generation of mathematicians, middle-aged mathematicians and also the middle and elementary school teachers. There is one person who deserves a special mention, and he is *Hua Luogeng*. He initiated and promoted the mathematical competition. He is also the author of the following books: Beyond *Yang hui*'s Triangle, Beyond the *pi* of *Zu Chongzhi*, Beyond the Magic Computation of *Sun-zi*, Mathematical Induction, and Mathematical Problems of Bee Hive. These were his books derived from mathematics competitions. When China resumed mathematical competition in 1978, he participated in setting problems and giving critique to solutions of the problems. Other outstanding books derived from the Chinese mathematics competitions are: Symmetry by *Duan Xuefu*, Lattice and Area by *Min Sihe*, One Stroke Drawing and Postman Problem by *Jiang Boju*.

After 1980, the younger mathematicians in China had taken over from the older generation of mathematicians in running the mathematical competition. They worked and strived hard to bring the level of mathematical competition in China to a new height. *Qiu Zonghu* is one such outstanding representative. From the training of contestants and leading the team 3 times to IMO to the organizing of the 31st IMO in China, he had contributed prominently and was awarded the *P. Erdős* prize.

Preparation for IMO

Currently, the selection process of participants for IMO in China is as follows.

First, the China Mathematical Competition, a national competition for high Schools, is organized on the second Sunday in September every year. The objectives are: to increase the interest of students in learning mathematics, to promote the development of co-curricular activities in mathematics, to help improve the teaching of mathematics in high schools, to discover and cultivate the talents and also to prepare for the IMO. This happens since 1981. Currently there are about 500,000 participants taking part.

Through the China Mathematical Competition, around 350 of students are selected to take part in the China Mathematical Olympiad or CMO, that is, the Winter Camp. The CMO lasts for 5 days and is held in November every year. The types and difficulties of the problems in CMO are very much similar to the IMO. There are also 3 problems to be completed within 4.5 hours each day. However, the score for each problem is 21 marks which add up to 126 marks in total. Starting from 1990, the Winter Camp instituted the *Chern Shiing-Shen* Cup for team championship. In 1991, the Winter Camp was officially renamed as the China Mathematical Olympiad (CMO). It is similar to the highest national mathematical competition in the former Soviet Union and the United States.

The CMO awards the first, second and third prizes. Among the participants of CMO, about 60 students are selected to participate in the training for IMO. The training takes place in March every year. After 6 to 8 tests and another 2 rounds of qualifying examinations, only 6 contestants are short-listed to form the China IMO national team to take part in the IMO in July.

Besides the China Mathematical Competition (for high schools), the Junior Middle School Mathematical Competition is also developing well. Starting from 1984, the competition is organized in April every year by the Popularization Committee of the Chinese Mathematical Society. The various provinces, cities and autonomous regions would rotate to host the event. Another mathematical competition for the junior middle schools is also conducted in April every year by the Middle School Mathematics Education Society of the Chinese Educational Society since 1998 till now.

The *Hua Luogeng* Gold Cup, a competition by invitation, had also been successfully conducted since 1986. The participating students comprise elementary six and junior middle one students. The format of the competition consists of a preliminary round, semi-finals in various provinces, cities, and autonomous regions, then the finals.

Mathematical competition in China provides a platform for students to showcase their talents in mathematics. It encourages learning of mathematics among students. It helps identify talented students and to provide them with differentiated learning opportunity. It develops co-curricular activities in mathematics. Finally, it brings about changes in the teaching of mathematics.

Contents

China Mathematical Competition

2016

There are 8 short-answer questions and 3 word problems, which should be solved in 120 minutes with a full score of 120 marks, in the first round test of 2016 China Mathematical Competition (CMC). While the scope of these questions and problems does not exceed the teaching requirements and content stipulated in the "Mathematics Syllabus for Full-time Ordinary Senior Middle Schools" issued by the Ministry of Education in 2000, smarter methods are needed to deal with them, so that contestants' mastery of basic knowledge and skills, as well as their abilities to comprehensively and flexibly apply these knowledge and skills in practice is examined.

There are 4 word problems (including one in plane geometry), which should be solved in 150 minutes with a full score of 180 marks, in the second round test (Complementary Test) of CMC. Their scope is in line with the International Mathematical Olympiad, plus some content of the Mathematical Competition Syllabus.

Part I Short-Answer Questions (Questions 1–8, eight marks each)

1　Given real number a satisfies $a < 9a^3 - 11a < |a|$, the range of a is _____.

Solution　From $a < |a|$ we have $a < 0$. Then the original inequalities become

$$1 > \frac{9a^3 - 11a}{a} > \frac{|a|}{a} = -1,$$

or $-1 < 9a^2 - 11 < 1$. Therefore $a^2 \in \left(\dfrac{10}{9}, \dfrac{4}{3}\right)$. Since $a < 0$ we have

$$a \in \left(-\frac{2\sqrt{3}}{3}, -\frac{\sqrt{10}}{3}\right).$$ □

2　Suppose complex numbers z, w satisfy

$$|z| = 3 \text{ and } (z + \overline{w})(\overline{z} - w) = 7 + 4i,$$

where i is the imaginary unit, and $\overline{z}$, $\overline{w}$ are conjugates of z, w, respectively. Then the modulus of $(z + 2\overline{w})(\overline{z} - 2w)$ is _____.

Solution　We have

$$7 + 4i = (z + \overline{w})(\overline{z} - w) = |z|^2 - |w|^2 - (zw - \overline{zw}).$$

Since $|z|^2$ and $|w|^2$ are real, and Re $(zw - \overline{zw}) = 0$, then

$$|z|^2 - |w|^2 = 7 \text{ and } zw - \overline{zw} = -4i.$$

As $|z| = 3$, $|w|^2 = 2$. Then

$$(z + 2\overline{w})(\overline{z} - 2w) = |z|^2 - 4|w|^2 - 2(zw - \overline{zw})$$
$$= 9 - 8 + 8i = 1 + 8i.$$

Finally, we get the modulus of $(z + 2\overline{w})(\overline{z} - 2w)$ is $\sqrt{1^2 + 8^2} = \sqrt{65}$. □

3　None of three positive real numbers u, v, w being 1, if

$$\log_u vw + \log_v w = 5 \text{ and } \log_v u + \log_w v = 3,$$

then the value of $\log_w u$ is _____.

Solution Let $\log_u v = a$, $\log_v w = b$. Then

$$\log_v u = \frac{1}{a}, \ \log_w v = \frac{1}{b}, \ \log_u vw = \log_u v + \log_u v \cdot \log_v w = a + ab.$$

The given formulas become

$$a + ab + b = 5 \quad \text{and} \quad \frac{1}{a} + \frac{1}{b} = 3,$$

from which we have $ab = \frac{5}{4}$. Therefore

$$\log_w u = \log_w v \cdot \log_v u = \frac{1}{ab} = \frac{4}{5}. \qquad \square$$

4 Bag A contains two 10-yuan banknotes and three one-yuan banknotes, and bag B contains four 5-yuan banknotes and three one-yuan banknotes. Now two banknotes are randomly taken out of each of the two bags. The probability that the sum of the face value of the remaining banknotes in A is greater than the sum of the face value of the remaining banknotes in B is _____.

Solution Define a as the sum of the face value of the two banknotes taken out of A, and b as that of the two banknotes taken out of B. Equivalently, we just need to find the probability that $a < b$. Since $b \le 5 + 5 = 10$, the two banknotes taken out of A must be both one-yuan, so there are $C_3^2 = 3$ choices. Furthermore, from $b > a = 2$ we know the two banknotes taken out of B cannot be both one-yuan, so there are $C_7^2 - C_3^2 = 18$ choices. Therefore, the required probability is

$$\frac{3 \times 18}{C_5^2 \times C_7^2} = \frac{54}{10 \times 21} = \frac{9}{35}. \qquad \square$$

5 Let P be the vertex of a cone with points A, B, C on the circumference of its bottom surface, satisfying $\angle ABC = 90°$, and M be the midpoint of AP. If $AB = 1$, $AC = 2$, $AP = \sqrt{2}$, then the dihedral angle M-BC-A is _____.

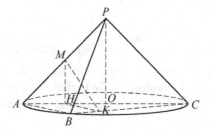

Solution From $\angle ABC = 90°$ we know AC is the diameter of the bottom circle, whose center is denoted by O. Then $PO \perp$ plane ABC. Since $AO = \frac{1}{2}AC = 1$, we have $PO = \sqrt{AP^2 - AO^2} = 1$.

Let H be the projection of M on the bottom surface. Then H is the midpoint of AO. If $HK \perp BC$ is taken at point K in the bottom surface, then $MK \perp BC$ is known from the three perpendicular theorem, so $\angle MKH$ equals the dihedral angle M-BC-A.

Since $MH = AH = \frac{1}{2}$ and $HK // AB$, we have $\dfrac{HK}{AB} = \dfrac{HC}{AC} = \dfrac{3}{4}$, i.e., $HK = \dfrac{3}{4}$. Then $\tan\angle MKH = \dfrac{MH}{HK} = \dfrac{2}{3}$, which means the dihedral angle M-BC-A is $\arctan\dfrac{2}{3}$. $\qquad\qquad\square$

6 Given $f(x) = \left(\sin\dfrac{kx}{10}\right)^4 + \left(\cos\dfrac{kx}{10}\right)^4$, where k is a positive integer. If for any real number a,

$$\{f(x) \mid a < x < a+1\} = \{f(x) \mid x \in \mathbb{R}\},$$

then the minimum of k is _____.

Solution We have

$$f(x) = \left(\sin^2\frac{kx}{10} + \cos^2\frac{kx}{10}\right)^2 - 2\sin^2\frac{kx}{10}\cos^2\frac{kx}{10}$$

$$= 1 - \frac{1}{2}\sin^2\frac{kx}{5} = \frac{1}{4}\cos\frac{2kx}{5} + \frac{3}{4}.$$

$f(x)$ reaches the maximum if and only if $x = \dfrac{5m\pi}{k} (m \in \mathbb{Z})$. Since any open interval $(a, a+1)$ with length 1 contains at least one maximum point, so $k > 5\pi$.

On the other hand, when $k > 5\pi$, any open interval $(a, a+1)$ will contains an entire period of $f(x)$. Therefore, the minimum of k is $[5\pi] + 1 = 16$. $\qquad\qquad\square$

7 Let hyperbola C: $x^2 - \dfrac{y^2}{3} = 1$, with the left and right focal points being F_1, and F_2, respectively. Crossing F_2 draw a line that intersects with the right half of the hyperbola at points P and Q, making $\angle F_1PQ = 90°$. Then the inscribed circle radius of $\triangle F_1PQ$ is

_____.

Solution By the property of a hyperbola, we know

$$F_1F_2 = 2 \times \sqrt{1+3} = 4 \text{ and } PF_1 - PF_2 = QF_1 - QF_2 = 2.$$

Since $\angle F_1PQ = 90°$, then $PF_1^2 + PF_2^2 = F_1F_2^2$. We have

$$PF_1 + PF_{2v} = \sqrt{2(PF_1^2 + PF_2^2) - (PF_1 - PF_2)^2}$$

$$= \sqrt{2 \times 4^2 - 2^2} = 2\sqrt{7}.$$

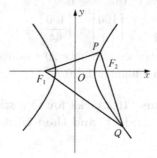

Therefore, the inscribed circle radius of $\triangle F_1PQ$ is

$$r = \frac{1}{2}(F_1P + PQ - F_1Q) = \frac{1}{2}(PF_1 + PF_2) - \frac{1}{2}(QF_1 - QF_2)$$

$$= \sqrt{7} - 1. \qquad \square$$

8 Let a_1, a_2, a_3, a_4 be four different integers among $1, 2, \ldots, 100$, satisfying

$$(a_1^2 + a_2^2 + a_3^2)(a_2^2 + a_3^2 + a_4^2) = (a_1a_2 + a_2a_3 + a_3a_4)^2.$$

Then the number of such ordered sequence $(a_1, \ a_2, \ a_3, \ a_4)$ is

_____ .

Solution By Cauchy inequality, we know

$$(a_1^2 + a_2^2 + a_3^2)(a_2^2 + a_3^2 + a_4^2) \ge (a_1a_2 + a_2a_3 + a_3a_4)^2.$$

The equality holds if and only if $\dfrac{a_1}{a_2} = \dfrac{a_2}{a_3} = \dfrac{a_3}{a_4}$, so the original question is equivalent to: find the number of ordered $(a_1, \ a_2, \ a_3, \ a_4)$, such that

$$\{a_1, a_2, a_3, a_4\} \subseteq \{1, 2, 3, \ldots, 100\}$$

and a_1, a_2, a_3, a_4 is a geometric sequence. Now suppose the common ratio of the sequence is $q \ne 1$, where q is a rational number. Then we can write $q = \dfrac{n}{m}$, where m, n are positive integers satisfying $m \ne n$ and $(m, n) = 1$.

When $n > m$, $a_4 = a_1 \cdot \left(\dfrac{n}{m}\right)^3 = \dfrac{a_1 n^3}{m^3}$. Then $l = \dfrac{a_1}{m^3}$ is a positive integer, as $(m^3, n^3) = 1$. Therefore, a_1, a_2, a_3, a_4 are $m^3 l$, $m^2 nl$, $mn^2 l$, $n^3 l$, respectively. Obviously, n can be 2, 3 or 4 only, as $5^3 > 100$; and for any n, the value of l can be any of $1, 2, \ldots, \left[\dfrac{100}{n^3}\right]$; furthermore, it is easy to see the value of q can only be 2, 3, $\dfrac{3}{2}$, 4 and $\dfrac{4}{3}$. Therefore, the number of required ordered arrays in this case is

$$\left[\frac{100}{8}\right] + \left[\frac{100}{27}\right] + \left[\frac{100}{27}\right] + \left[\frac{100}{64}\right] + \left[\frac{100}{64}\right] = 12 + 3 + 3 + 1 + 1 = 20.$$

When $n < m$, by symmetry we also have 20 required ordered arrays.

Therefore, the required number is 40.   □

Part II Word Problems (16 marks for Question 9, 20 marks each for Questions 10 and 11, and then 56 marks in total)

9   In $\triangle ABC$, we know

$$\overrightarrow{AB} \cdot \overrightarrow{AC} + 2\overrightarrow{BA} \cdot \overrightarrow{BC} = 3\overrightarrow{CA} \cdot \overrightarrow{CB}.$$

Find the maximum value of $\sin C$.

Solution   By the definition of scalar products and the law of cosines, we know

$$\overrightarrow{AB} \cdot \overrightarrow{AC} = cb \cos A = \frac{b^2 + c^2 - a^2}{2}.$$

In the same way,

$$\overrightarrow{BA} \cdot \overrightarrow{BC} = \frac{a^2 + c^2 - b^2}{2} \quad \text{and} \quad \overrightarrow{CA} \cdot \overrightarrow{CB} = \frac{a^2 + b^2 - c^2}{2}.$$

Then the original equation becomes

$$b^2 + c^2 - a^2 + 2(a^2 + c^2 - b^2) = 3(a^2 + b^2 - c^2),$$

or $a^2 + 2b^2 = 3c^2$. So we have

$$\cos C = \frac{a^2 + b^2 - c^2}{2ab} = \frac{a^2 + b^2 - \dfrac{1}{3}(a^2 + 2b^2)}{2ab}$$

$$= \frac{a}{3b} + \frac{b}{6a} \geq 2\sqrt{\frac{a}{3b} \cdot \frac{b}{6a}} = \frac{\sqrt{2}}{3}.$$

Then $\sin C = \sqrt{1 - \cos^2 C} \leq \dfrac{\sqrt{7}}{3}$, and the equality holds if and only if $a\colon b\colon c = \sqrt{3}\colon \sqrt{6}\colon \sqrt{5}$. Therefore the maximum value of $\sin C$ is $\dfrac{\sqrt{7}}{3}$.   □

10 Given that $f(x)$ is an odd function on $\mathbb{R}$, $f(1) = 1$, and $f\left(\dfrac{x}{x-1}\right) = xf(x)$ for any $x < 0$, find the value of

$$f(1)f\left(\frac{1}{100}\right) + f\left(\frac{1}{2}\right)f\left(\frac{1}{99}\right) + f\left(\frac{1}{3}\right)f\left(\frac{1}{98}\right)$$

$$+ \cdots + f\left(\frac{1}{50}\right)f\left(\frac{1}{51}\right).$$

Solution Define $a_n = f\left(\dfrac{1}{n}\right)$ $(n = 1, 2, 3, \ldots)$; then $a_1 = f(1) = 1$. For $x = -\dfrac{1}{k}(k \in \mathbb{N}^+)$, Note that

$$\frac{x}{x-1} = \frac{-\dfrac{1}{k}}{-\dfrac{1}{k} - 1} = \frac{1}{k+1}, \text{ and } f(x) \text{ is an odd function.}$$

We have

$$f\left(\frac{1}{k+1}\right) = -\frac{1}{k} \cdot f\left(-\frac{1}{k}\right) = \frac{1}{k} \cdot f\left(\frac{1}{k}\right),$$

and that means $\dfrac{a_{k+1}}{a_k} = \dfrac{1}{k}$. Therefore,

$$a_n = a_1 \cdot \sum_{k=1}^{n-1} \frac{a_{k+1}}{a_k} = \sum_{k=1}^{n-1} \frac{1}{k} = \frac{1}{(n-1)!}.$$

Then

$$\sum_{i=1}^{50} a_i a_{101-i} = \sum_{i=1}^{50} \frac{1}{(i-1)! \cdot (100-i)!} = \sum_{i=0}^{49} \frac{1}{i! \cdot (99-i)!}$$

$$= \frac{1}{99!} \cdot \sum_{i=0}^{49} C_{99}^i = \frac{1}{99!} \cdot \sum_{i=0}^{49} \frac{1}{2}(C_{99}^i + C_{99}^{99-i})$$

$$= \frac{1}{99!} \times \frac{1}{2} \times 2^{99} = \frac{2^{98}}{99!}. \qquad \square$$

11 As shown in Fig. 11.1, in the plane rectangular coordinate system xOy, point F on the positive semi-axis of the x-axis is the focus and O is the vertex of parabola C. Through P on C in the first quadrant and Q on the negative semi-axis of the x-axis, draw line PQ tangent

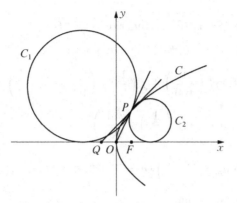

Fig. 11.1

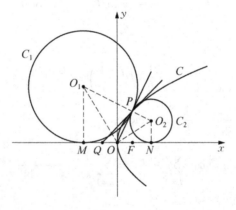

Fig. 11.2

to C, and $|PQ| = 2$. Circles C_1 and C_2 are tangent to line OP at P and are both tangent to the x-axis. Find the coordinates of F so that the sum of the areas of circles C_1 and C_2 takes the minimum value.

Solution   Suppose the equation of parabola C is $y^2 = 2px(p > 0)$, $Q(-a, 0)$ $(a > 0)$, and $O_1(x_1, y_1)$, $O_2(x_2, y_2)$ are the centers of C_1, C_2, respectively. Let the equation of line PQ be $x = my - a$ $(m > 0)$. Combining it with the parabola equation to eliminate x, we get

$$y^2 - 2pmy + 2pa = 0.$$

As PQ is tangent to C at point P, so

$$\Delta = 4p^2m^2 - 4 \cdot 2pa = 0.$$

We get $m = \sqrt{\dfrac{2a}{p}}$, and then $P(x_P, y_P) = (a, \sqrt{2pa})$.

Therefore,

$$|PQ| = \sqrt{1 + m^2} \cdot |y_P - 0|$$
$$= \sqrt{1 + \frac{2a}{p}} \cdot \sqrt{2pa}$$
$$= \sqrt{2a(p + 2a)} = 2.$$

So

$$4a^2 + 2pa = 4. \tag{1}$$

We have $OP \perp O_1O_2$, as OP is tangent to C_1, C_2 at P. Suppose C_1, C_2 are tangent to the x-axis at M, N, respectively. Then OO_1 and OO_2 are bisectors of $\angle POM$ and $\angle PON$, respectively, so $\angle O_1OO_2 = 90°$. From the projective theorem, we have

$$y_1 \cdot y_2 = O_1M \cdot O_2N$$
$$= O_1P \cdot O_2P = OP^2$$
$$= x_P^2 + y_P^2 = a^2 + 2pa.$$

By (1), we then get

$$y_1 \cdot y_2 = a^2 + 2pa = 4 - 3a^2. \tag{2}$$

Since O_1, P, O_2 are collinear, so

$$\frac{y_1 - \sqrt{2pa}}{\sqrt{2pa} - y_2} = \frac{y_1 - y_P}{y_P - y_2} = \frac{O_1P}{PO_2} = \frac{O_1M}{O_2N} = \frac{y_1}{y_2},$$

from which we have

$$y_1 + y_2 = \frac{2}{\sqrt{2pa}} \cdot y_1y_2. \tag{3}$$

Let $T = y_1^2 + y_2^2$. Then the sum of the areas of circles C_1 and C_2 is πT. From (2), (3),

$$T = (y_1 + y_2)^2 - 2y_1y_2 = \frac{4}{2pa}y_1^2y_2^2 - 2y_1y_2$$
$$= \frac{4}{4 - 4a^2}(4 - 3a^2)^2 - 2(4 - 3a^2)$$
$$= \frac{(4 - 3a^2)(2 - a^2)}{1 - a^2}.$$

Let $t = 1 - a^2$. Then from $4t = 4 - 4a^2 = 2pa > 0$, we know $t > 0$. Therefore,

$$T = \frac{(3t+1)(t+1)}{t} = 3t + \frac{1}{t} + 4$$

$$\geq 2\sqrt{3t \cdot \frac{1}{t}} + 4 = 2\sqrt{3} + 4.$$

The equality holds if and only if $t = \dfrac{\sqrt{3}}{3}$. Then $a = \sqrt{1-t} = \sqrt{\left(1 - \dfrac{1}{\sqrt{3}}\right)}$, when πT reaches the minimum. Then by ① we have

$$\frac{p}{2} = \frac{1 - a^2}{a} = \frac{t}{\sqrt{1 - \dfrac{1}{\sqrt{3}}}} = \frac{\sqrt{3}t}{\sqrt{3 - \sqrt{3}}} = \frac{1}{\sqrt{3 - \sqrt{3}}}.$$

Therefore the coordinates of F is $\left(\dfrac{1}{\sqrt{3 - \sqrt{3}}}, 0\right)$. □

China Mathematical Competition

2017

While the scope of the test questions in the first round of the 2017 China Mathematical Competition does not exceed the teaching requirements and content specified in the "General High School Mathematics Curriculum Standards (Experiments)" promulgated by the Ministry of Education of China in 2003, the methods of proposing the questions have been improved. The emphasis is placed on testing the students' basic knowledge and skills, and their abilities to integrate and flexibly use them. Each test paper includes eight fill-in-the-blank questions and three answer questions. The answer time is 80 minutes, and the full score is 120 points.

The scope of the test questions in the second round (Complementary Test) is in line with the International Mathematical Olympiad, with this some expanded knowledge, plus a few contents of the Mathematical Competition Syllabus. Each test paper consists of four answer questions, including a plane geometry one, and the answering time is 150 minutes. The full score is 180 points.

Test Paper A, the First Round

Part I Short-Answer Questions (Questions 1-8, eight marks each)

① Let $f(x)$ be defined on $\mathbb{R}$, satisfying, for each real number x,

$$f(x+3) \cdot f(x-4) = -1.$$

Moreover, when $0 \leq x < 7, f(x) = \log_2(9-x)$. Then $f(-100)$ equals

_____.

Solution Clearly, $f(x+14) = -\dfrac{1}{f(x+7)} = f(x)$, and it follows that

$$f(-100) = f(-100 + 14 \times 7) = f(-2) = -\frac{1}{f(5)} = -\frac{1}{\log_2 4} = -\frac{1}{2}. \quad \square$$

② Let x, y be real numbers such that $x^2 + 2\cos y = 1$. The range of $x - \cos y$ is _____.

Solution Since

$$x^2 = 1 - 2\cos y \in [-1, 3],$$

it follows that $x \in [-\sqrt{3}, \sqrt{3}]$. By $\cos y = \dfrac{1-x^2}{2}$, we have

$$x - \cos y = x - \frac{1-x^2}{2} = \frac{1}{2}(x+1)^2 - 1.$$

When $x = -1$, $x - \cos y$ attains the minimum value -1 (as y equals $\dfrac{\pi}{2}$); when $x = \sqrt{3}$, $x - \cos y$ attains the maximum value $\sqrt{3}+1$ (as y equals π). Since the range of $\dfrac{1}{2}(x+1)^2 - 1$ is $[-1, \sqrt{3}+1]$, we find the range of $x - \cos y$ as $[-1, \sqrt{3}+1]$. $\quad \square$

③ In the Cartesian plane, ellipse C is given by $\dfrac{x^2}{9} + \dfrac{y^2}{10} = 1$, point F is the upper focus of C, A is the right vertex of C, and P is a moving point on C in the first quadrant. The maximum area of quadrilateral $OAPF$ is _____.

Solution Clearly, we have $A(3, 0)$ and $F(0, 1)$. Let the coordinates of P be

$$(3\cos\theta, \sqrt{10}\sin\theta), \theta \in \left(0, \frac{\pi}{2}\right).$$

Then

$$S_{OAPF} = S_{\triangle OAP} + S_{\triangle OFP} = \frac{1}{2} \cdot 3 \cdot \sqrt{10}\sin\theta + \frac{1}{2} \cdot 1 \cdot 3\cos\theta$$

$$= \frac{3}{2}(\sqrt{10}\sin\theta + \cos\theta) = \frac{3\sqrt{11}}{2}\sin(\theta + \varphi),$$

in which $\varphi = \arctan\dfrac{\sqrt{10}}{10}$.

When $\theta = \arctan\sqrt{10}$, quadrilateral $OAPF$ has the maximal area as $\dfrac{3\sqrt{11}}{2}$. $\qquad\square$

4 A three-digit positive integer is called a "stable number" if any two consecutive digits differ by at most 1. The number of stable numbers is _____.

Solution Let $\overline{abc}$ be a stable number.

If $b = 0$, then $a = 1$, $c \in \{0, 1\}$, and there are 2 stable numbers.

If $b = 1$, then $a \in \{1, 2\}$, $c \in \{0, 1, 2\}$, and there are $2 \times 3 = 6$ stable numbers.

If $2 \leq b \leq 8$, then $a, c \in \{b-1, b, b+1\}$, and there are $7 \times 3 \times 3 = 63$ stable numbers.

If $b = 9$, then $a, c \in \{8, 9\}$, and there are $2 \times 2 = 4$ stable numbers.

Therefore, the number of stable numbers is $2 + 6 + 63 + 4 = 75$. $\qquad\square$

5 In tetrahedron P-ABC, $AB = 1$, $AP = 2$. Plane α through AB cuts the tetrahedron into two solids with equal volumes. Then the cosine of the angle between line PC and plane α equals _____.

Solution Let K, M be the midpoints of AB, PC, respectively. Clearly, plane ABM is exactly plane α. By the Apollonius's theorem,

$$AM^2 = \frac{1}{2}(AP^2 + AC^2) - \frac{1}{4}PC^2 = \frac{1}{2}(2^2 + 1^2) - \frac{1}{4} \times 2^2 = \frac{3}{2},$$

and hence

$$KM = \sqrt{AM^2 - AK^2} = \sqrt{\frac{3}{2} - \left(\frac{1}{2}\right)^2} = \frac{\sqrt{5}}{2}.$$

Furthermore, MK is the projection of line PC on α, $MC = 1$, $KC = \dfrac{\sqrt{3}}{2}$. Therefore

$$\cos \angle KMC = \frac{KM^2 + MC^2 - KC^2}{2KM \cdot MC} = \frac{\dfrac{5}{4} + 1 - \dfrac{3}{4}}{\sqrt{5}} = \frac{3\sqrt{5}}{10},$$

i.e., the angle between line PC and plane α has cosine value $\dfrac{3\sqrt{5}}{10}$. □

6 A plane point set is given by $K = \{(x, y) \mid x, y = -1, 0, 1\}$. Choose 3 points from K at random. Then the probability that there are two points among them with distance $\sqrt{5}$ is _____.

Solution   There are 9 points in K, and the number of ways of choosing 3 points from K is $\mathrm{C}_9^3 = 84$.

Let the points be $A_1, A_2, \ldots, A_8$ and O, as shown in the figure below.

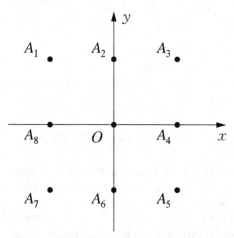

There are 8 pairs of points at distance $\sqrt{5}$. Let us choose one pair among them. By symmetry, we may assume the chosen pair are A_1 and A_4. Since there are 7 ways to choose the other point, there are $7 \times 8 = 56$ unordered triples. On the other hand, each $A_i (i = 1, 2, \ldots, 8)$ is at distance $\sqrt{5}$ from points A_{i+3}, A_{i+5} (indices modulo 8). This implies that there are 8 triples

$$\{A_i, A_{i+3}, A_{i+5}\} (i = 1, 2, \ldots, 8)$$

being counted twice. Hence the total number of unordered triples is $56 - 8 = 48$, and the probability is $\dfrac{48}{84} = \dfrac{4}{7}$. □

7 In $\triangle ABC$, M is the midpoint of BC, and N is the midpoint of BM. If $\angle A = \dfrac{\pi}{3}$, and the area of $\triangle ABC$ is $\sqrt{3}$, then the minimum value of $\overrightarrow{AM} \cdot \overrightarrow{AN}$ is _____.

Solution Clearly,

$$\overrightarrow{AM} = \frac{1}{2}(\overrightarrow{AB} + \overrightarrow{AC}), \overrightarrow{AN} = \frac{3}{4}\overrightarrow{AB} + \frac{1}{4}\overrightarrow{AC},$$

and it follows that

$$\overrightarrow{AM} \cdot \overrightarrow{AN} = \frac{1}{2}(\overrightarrow{AB} + \overrightarrow{AC}) \cdot \left(\frac{3}{4}\overrightarrow{AB} + \frac{1}{4}\overrightarrow{AC} \right)$$
$$= \frac{1}{8}(3|\overrightarrow{AB}|^2 + |\overrightarrow{AC}|^2 + 4\overrightarrow{AB} \cdot \overrightarrow{AC}).$$

Since

$$\sqrt{3} = S_{\triangle ABC} = \frac{1}{2} \cdot |\overrightarrow{AB}| \cdot |\overrightarrow{AC}| \cdot \sin A = \frac{\sqrt{3}}{4} \cdot |\overrightarrow{AB}| \cdot |\overrightarrow{AC}|,$$

we obtain $|\overrightarrow{AB}| \cdot |\overrightarrow{AC}| = 4$, and furthermore

$$\overrightarrow{AB} \cdot \overrightarrow{AC} = |\overrightarrow{AB}| \cdot |\overrightarrow{AC}| \cdot \cos A = 2.$$

Hence

$$\overrightarrow{AM} \cdot \overrightarrow{AN} \geq \frac{1}{8}(2\sqrt{3|\overrightarrow{AB}|^2 \cdot |\overrightarrow{AC}|^2} + 4\overrightarrow{AB} \cdot \overrightarrow{AC})$$

$$= \frac{\sqrt{3}}{4}|\overrightarrow{AB}| \cdot |\overrightarrow{AC}| + \frac{1}{2}\overrightarrow{AB} \cdot \overrightarrow{AC} = \sqrt{3} + 1.$$

When $|\overrightarrow{AB}| = \dfrac{2}{\sqrt[4]{3}}$, $|\overrightarrow{AC}| = 2 \times \sqrt[4]{3}$, $\overrightarrow{AM} \cdot \overrightarrow{AN}$ attains the minimum value as $\sqrt{3} + 1$. □

8 Let $\{a_n\}, \{b_n\}$ be strictly increasing positive integer sequences, satisfying: $a_{10} = b_{10} < 2017$, $a_{n+2} = a_{n+1} + a_n$, $b_{n+1} = 2b_n$ for every positive integer n. Then all possible values of $a_1 + b_1$ are _____.

Solution   Clearly, a_1, a_2, b_1 are positive integers, and $a_1 < a_2$.

Since $2017 > b_{10} = 2^9 \cdot b_1 = 512b_1$, $b_1 \in \{1, 2, 3\}$. Repeatedly use the recurrence formula of $\{a_n\}$ to derive

$$a_{10} = a_9 + a_8 = 2a_8 + a_7 = 3a_7 + 2a_6$$

$$= 5a_6 + 3a_5 = 8a_5 + 5a_4 = 13a_4 + 8a_3$$

$$= 21a_3 + 13a_2 = 34a_2 + 21a_1.$$

Thus

$$21a_1 \equiv a_{10} = b_{10} = 512b_1 \equiv 2b_1 \pmod{34}.$$

As $13 \times 21 = 34 \times 8 + 1$, we have

$$a_1 \equiv 13 \times 21a_1 \equiv 13 \times 2b_1 = 26b_1 \pmod{34}. \qquad ①$$

On the other hand, notice that $a_1 < a_2$, which implies $55a_1 < 34a_2 + 21a_1 = 512b_1$, or

$$a_1 < \frac{512}{55}b_1. \qquad ②$$

If $b_1 = 1$, ①, ② become $a_1 \equiv 26 \pmod{34}$, $a_1 < \dfrac{512}{55}$; no integer solution exists.

If $b_1 = 2$, ①, ② become $a_1 \equiv 52 \pmod{34}$, $a_1 < \dfrac{1024}{55}$; we have a unique solution $a_1 = 18$, and $a_1 + b_1 = 20$.

If $b_1 = 3$, ①, ② become $a_1 \equiv 78 \pmod{34}$, $a_1 < \dfrac{1536}{55}$; we have a unique solution $a_1 = 10$, and $a_1 + b_1 = 13$.

As a summary, $a_1 + b_1$ could be 13 or 20. $\square$

Part II Word problems (Questions 9–11, 56 marks in total for three questions)

    9   (16 marks) Let k, m be real numbers. If $|x^2 - kx - m| \leq 1$ holds for all $x \in [a, b]$, prove $b - a \leq 2\sqrt{2}$.

Solution   Let $f(x) = x^2 - kx - m$, $x \in [a, b]$. Then $f(x) \in [-1, 1]$, and in particular

$$f(a) = a^2 - ka - m \leq 1, \tag{1}$$

$$f(b) = b^2 - kb - m \leq 1, \tag{2}$$

$$f\left(\frac{a+b}{2}\right) = \left(\frac{a+b}{2}\right)^2 - k \cdot \frac{a+b}{2} - m \geq -1. \tag{3}$$

Take $(1) + (2) - 2 \times (3)$ to get

$$\frac{(a-b)^2}{2} = f(a) + f(b) - 2f\left(\frac{a+b}{2}\right) \leq 4.$$

Hence $b - a \leq 2\sqrt{2}$. □

10 (20 marks) Let x_1, x_2, x_3 be nonnegative real numberss satisfying $x_1 + x_2 + x_3 = 1$. Find the maximum and minimum values of

$$(x_1 + 3x_2 + 5x_3)\left(x_1 + \frac{x_2}{3} + \frac{x_3}{5}\right).$$

Solution   By Cauchy inequality, we have

$$(x_1 + 3x_2 + 5x_3)\left(x_1 + \frac{x_2}{3} + \frac{x_3}{5}\right)$$

$$\geq \left(\sqrt{x_1} \cdot \sqrt{x_1} + \sqrt{3x_2} \cdot \sqrt{\frac{x_2}{3}} + \sqrt{5x_3} \cdot \sqrt{\frac{x_3}{5}}\right)^2$$

$$= (x_1 + x_2 + x_3)^2 = 1,$$

where the equality holds at $x_1 = 1$, $x_2 = 0$, $x_3 = 0$. Therefore, the minimum value is 1.

On the other hand,

$$(x_1 + 3x_2 + 5x_3)\left(x_1 + \frac{x_2}{3} + \frac{x_3}{5}\right)$$

$$= \frac{1}{5}(x_1 + 3x_2 + 5x_3)\left(5x_1 + \frac{5x_2}{3} + x_3\right)$$

$$\leq \frac{1}{5} \cdot \frac{1}{4}\left[(x_1 + 3x_2 + 5x_3) + \left(5x_1 + \frac{5x_2}{3} + x_3\right)\right]^2$$

$$= \frac{1}{20}\left(6x_1 + \frac{14}{3}x_2 + 6x_3\right)^2$$

$$\leq \frac{1}{20}(6x_1 + 6x_2 + 6x_3)^2 = \frac{9}{5}.$$

The equalities hold simultaneously at $x_1 = \dfrac{1}{2}$, $x_2 = 0$, $x_3 = \dfrac{1}{2}$. So the maximum value is $\dfrac{9}{5}$. $\qquad\qquad\square$

11 (20 marks) Let complex numbers z_1, z_2 satisfy $\text{Re}(z_1) > 0$, $\text{Re}(z_2) > 0$, and $\text{Re}(z_1^2) = \text{Re}(z_2^2) = 2$ ($\text{Re}(z)$ represents the real part of z).

(1) Find the minimum of $\text{Re}(z_1 z_2)$;

(2) Find the minimum of $|z_1 + 2| + |\overline{z_2} + 2| - |\overline{z_1} - z_2|$.

Solution (1) For $k = 1, 2$, let $z_k = x_k + y_k \text{i}$ (x_k, $y_k \in \mathbb{R}$). Clearly,

$$x_k = \text{Re}(z_k) > 0, x_k^2 - y_k^2 = \text{Re}(z_k^2) = 2.$$

This implies that

$$\text{Re}(z_1 z_2) = \text{Re}((x_1 + y_1\text{i})(x_2 + y_2\text{i})) = x_1 x_2 - y_1 y_2$$

$$= \sqrt{(y_1^2 + 2)(y_2^2 + 2)} - y_1 y_2$$

$$\geq (|y_1 y_2| + 2) - y_1 y_2 \geq 2.$$

When $z_1 = z_2 = \sqrt{2}$, $\text{Re}(z_1 z_2) = 2$. Hence the minimum of $\text{Re}(z_1 z_2)$ is 2.

(2) In the Cartesian plane, let complex number z_k correspond to point $P_k(x_k, y_k)$, for $k = 1, 2$. Let P_2' and P_2 be symmetric about the x-axis. Then P_1, P_2' are on the right branch of the hyperbola C: $x^2 - y^2 = 2$.

Let F_1, F_2 be the foci of C. Evidently, $F_1(-2, 0), F_2(2, 0)$. By definition of hyperbola,

$$|P_1 F_1| = |P_1 F_2| + 2\sqrt{2}, |P_2' F_1| = |P_2' F_2| + 2\sqrt{2},$$

and furthermore,

$$|z_1 + 2| + |\overline{z_2} + 2| - |\overline{z_1} - z_2|$$

$$= |z_1 + 2| + |\overline{z_2} + 2| - |z_1 - \overline{z_2}|$$

$$= |P_1 F_1| + |P_2' F_1| - |P_1 P_2'|$$

$$= 4\sqrt{2} + |P_1 F_2| + |P_2' F_2| - |P_1 P_2'|$$

$$\geq 4\sqrt{2},$$

where the quality holds only when F_2 lies on the segment $P_1 P_2'$ (for example, $z_1 = z_2 = 2 + \sqrt{2}\text{i}$, and F_2 is the midpoint of $P_1 P_2'$).

Thus the minimum of $|z_1 + 2| + |\overline{z_2} + 2| - |\overline{z_1} - z_2|$ is $4\sqrt{2}$. $\qquad\square$

Test Paper B, the First Round

Part I Short-Answer Questions (Questions 1–8, eight marks each)

1. Suppose a geometric sequence $\{a_n\}$ satisfies $a_2 = \sqrt{2}$, $a_3 = \sqrt[3]{3}$. Then $\dfrac{a_1 + a_{2011}}{a_7 + a_{2017}}$ equals _____.

Solution The common ratio of $\{a_n\}$ is $q = \dfrac{a_3}{a_2} = \dfrac{\sqrt[3]{3}}{\sqrt{2}}$, and hence

$$\frac{a_1 + a_{2011}}{a_7 + a_{2017}} = \frac{a_1 + a_{2011}}{q^6(a_1 + a_{2011})} = \frac{1}{q^6} = \frac{8}{9}. \qquad \square$$

2. Let complex number z satisfy the equation $z + 9 = 10\bar{z} + 22i$. Then $|z|$ equals _____.

Solution Let $z = a + bi$, a, $b \in \mathbb{R}$. Clearly,

$$(a + 9) + bi = 10a + (-10b + 22)i.$$

Comparing the real and the imaginary parts, we have $\begin{cases} a + 9 = 10a, \\ b = -10b + 22, \end{cases}$ which gives $a = 1$, $b = 2$, $z = 1 + 2i$. So $|z| = \sqrt{5}$. $\qquad \square$

3. Let $f(x)$ be defined on $\mathbb{R}$, such that $f(x) + x^2$ is an odd function, and $f(x) + 2^x$ is an even function. Then $f(1)$ equals _____.

Solution By definition of odd and even functions, we have

$$f(1) + 1 = -[f(-1) + (-1)^2] = -f(-1) - 1,$$

$$f(1) + 2 = f(-1) + \frac{1}{2}.$$

Adding the above equations to cancel $f(-1)$, it follows $2f(1) + 3 = -\dfrac{1}{2}$, or $f(1) = -\dfrac{7}{4}$. $\qquad \square$

4. In $\triangle ABC$, $\sin A = 2 \sin C$, and the side lengths a, b, c form a geometric progression. Then $\cos A$ equals _____.

Solution   By the law of sines, $\dfrac{a}{c} = \dfrac{\sin A}{\sin C} = 2$. From $b^2 = ac$, we have $a :$
$b : c = 2 : \sqrt{2} : 1$. Then by the law of cosines,

$$\cos A = \frac{b^2 + c^2 - a^2}{2bc} = \frac{(\sqrt{2})^2 + 1^2 - 2^2}{2 \times \sqrt{2} \times 1} = -\frac{\sqrt{2}}{4}. \qquad \square$$

5   Given regular tetrahedron $ABCD$, points E, F are on AB, AC, respectively, such that $BE = 3$, $EF = 4$, and EF is parallel to plane BCD. Then the area of $\triangle DEF$ is _____.

Solution   Clearly, EF is parallel to BC. Since all the faces of $ABCD$ are equilateral triangles, we have

$$AE = AF = EF = 4, AD = AB = AE + BE = 7.$$

By the law of cosines,

$$DE = \sqrt{AD^2 + AE^2 - 2AD \cdot AE \cdot \cos 60°}$$
$$= \sqrt{49 + 16 - 28} = \sqrt{37},$$

and $DF = \sqrt{37}$ likewise.

     Let DH be an altitude of isosceles $\triangle DEF$, H being the foot on EF, as shown in the figure.

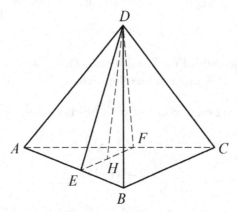

Then $EH = \dfrac{1}{2} EF = 2$, and

$$DH = \sqrt{DE^2 - EH^2} = \sqrt{33}.$$

We conclude $S_{\triangle DEF} = \dfrac{1}{2} \cdot EF \cdot DH = 2\sqrt{33}$. $\square$

6 In the Cartesian plane, let set $K = \{(x, y) \mid x, y = -1, 0, 1\}$. Suppose that three points are chosen from K at random. Then the probability of no pairwise distance exceeding 2 equals _____.

Solution The number of ways to choose 3 points from $|K| = 9$ is $C_9^3 = 84$. If no pairwise distance is greater than 2, there are three situations:

(1) The three points are collinear; 6 cases.
(2) The three points are the vertices of a right isosceles triangle with side lengths 1, 1, $\sqrt{2}$; $4 \times 4 = 16$ cases.
(3) The three points are the vertices of a right isosceles triangle with side lengths $\sqrt{2}$, $\sqrt{2}$, 2. When the right-angle vertex is at $(0, 0)$: 4 cases; when the right-angle vertex is at $(\pm 1, 0)$, $(0, \pm 1)$: 4 cases.

There are $6 + 16 + 8 = 30$ cases in total, and hence the probability is $\dfrac{30}{84} = \dfrac{5}{14}$. □

7 Let $a \neq 0$ be a real number. The foci of the quadric curve $x^2 + ay^2 + a^2 = 0$ are at distance 4 from each other. Then a is _____.

Solution Rewrite the equation as $-\dfrac{x^2}{a^2} - \dfrac{y^2}{a} = 1$. Clearly, $-a > 0$, and the curve is a hyperbola, represented by $\dfrac{y^2}{(\sqrt{-a})^2} - \dfrac{x^2}{(-a)^2} = 1$. We have

$$c^2 = (\sqrt{-a})^2 + (-a)^2 = a^2 - a,$$

and by $2c = 4$, $a^2 - a = 4$. Since $a < 0$, $a = \dfrac{1 - \sqrt{17}}{2}$. □

8 Suppose positive integers a, b, c satisfy $2017 \geq 10a \geq 100b \geq 1000c$. The number of such (a, b, c) triples is _____.

Solution From the given inequality, we have $c \leq \left[\dfrac{2017}{1000}\right] = 2$.

If $c = 1$, $10 \leq b \leq 20$. For each b, due to the restriction $10b \leq a \leq 201$, there are $202 - 10b$ of a's, and thus the number of triples is

$$\sum_{b=10}^{20}(202 - 10b) = \dfrac{(102 + 2) \times 11}{2} = 572.$$

If $c = 2$, $20 \leq b \leq \left[\dfrac{2017}{100}\right]$, $b = 20$. Then from $200 \leq a \leq \left[\dfrac{2017}{10}\right] = 201$, we have $a = 200, 201$. There are two triples (a, b, c).

In summary, there are $572 + 2 = 574$ triples. □

Part II Word problems (Questions 9–11, 56 marks in total for three questions)

> **9** (16 marks) Given that the inequality $|2^x - a| < |5 - 2^x|$ holds for all $x \in [1, 2]$, find the range of real number a.

Solution Let $t = 2^x$. Clearly, $t \in [2, 4]$, and $|t - a| < |5 - t|$ holds for all $t \in [2, 4]$. We eliminate the absolute value sign by taking squares on both sides:

$$|t - a| < |5 - t| \Leftrightarrow (t - a)^2 < (5 - t)^2$$
$$\Leftrightarrow (2t - a - 5)(5 - a) < 0.$$

For given real a, let $f(t) = (2t - a - 5)(5 - a)$. Notice that $f(t)$ is a linear function of t (or constant 0 when $a = 5$). To have $f(t) < 0$ for all $t \in [2, 4]$, it suffices to require that

$$\begin{cases} f(2) = (-1 - a)(5 - a) < 0, \\ f(4) = (3 - a)(5 - a) < 0, \end{cases}$$

which gives $3 < a < 5$.

Therefore, the range of a is $3 < a < 5$. $\qquad\square$

> **10** Given an arithmetic progression $\{a_n\}$, let progression $\{b_n\}$ satisfy $b_n = a_{n+1}a_{n+2} - a_n^2$, for $n = 1, 2, \ldots$.
>
> (1) Show that $\{b_n\}$ is also an arithmetic progression.
> (2) Let $d \neq 0$ be the common difference of both $\{a_n\}$ and of $\{b_n\}$. Suppose there exist positive integers s, t, such that $a_s + b_t$ is an integer. Find the minimum value of $|a_1|$.

Solution (1) Let d be the common difference of $\{a_n\}$. A straightforward calculation yields

$$b_{n+1} - b_n = (a_{n+2}a_{n+3} - a_{n+1}^2) - (a_{n+1}a_{n+2} - a_n^2)$$
$$= a_{n+2}(a_{n+3} - a_{n+1}) - (a_{n+1} + a_n)(a_{n+1} - a_n)$$
$$= a_{n+2} \cdot 2d - (a_{n+1} + a_n) \cdot d$$
$$= (2a_{n+2} - a_{n+1} - a_n) \cdot d = 3d^2.$$

Hence $\{b_n\}$ is also an arithmetic progression.

(2) By the result in (1), $3d^2 = d$, $d = \dfrac{1}{3}$ since $d \neq 0$. We have

$$b_n = a_{n+1}a_{n+2} - a_n^2 = (a_n + d)(a_n + 2d) - a_n^2$$

$$= 3da_n + 2d^2 = a_n + \frac{2}{9}.$$

If $a_s + b_t \in \mathbb{Z}$, then

$$a_s + b_t = a_s + a_t + \frac{2}{9} = a_1 + (s-1)d + a_1 + (t-1)d + \frac{2}{9}$$

$$= 2a_1 + \frac{s+t-2}{3} + \frac{2}{9} \in \mathbb{Z}.$$

Denote $l = 2a_1 + \dfrac{s+t-2}{3} + \dfrac{2}{9}$, $l \in \mathbb{Z}$. For integers s, t, $18a_1 = 3(3l - s - t + 1) + 1$ is a nonzero integer, thus $|18a_1| \geq 1$, $|a_1| \geq \dfrac{1}{18}$.

When $a_1 = \dfrac{1}{18}$, $a_1 + b_3 = \dfrac{1}{18} + \dfrac{17}{18} = 1 \in \mathbb{Z}$.

Therefore, the minimum value of $|a_1|$ is $\dfrac{1}{18}$. $\qquad \square$

11 Let C_1: $y^2 = 4x$, C_2: $(x-4)^2 + y^2 = 8$ be curves in the Cartesian plane. Through point P on C_1 draw line l with 45-degree inclination that intersects C_2 at distinct points Q, R. Find the range of $|PQ| \cdot |PR|$.

Solution Let $P = (t^2, 2t)$, and we obtain the equation of l as $y = x + 2t - t^2$. Plug it into the equation for C_2,

$$(x-4)^2 + (x + 2t - t^2)^2 = 8,$$

and simplify, to get

$$2x^2 - 2(t^2 - 2t + 4)x + (t^2 - 2t)^2 + 8 = 0. \qquad \textcircled{1}$$

Since Q, R are distinct intersections, the discriminant of quadratic equation $\textcircled{1}$ is positive, so

$$\frac{\Delta}{4} = (t^2 - 2t + 4)^2 - 2[(t^2 - 2t)^2 + 8]$$

$$= (t^2 - 2t)^2 - 8(t^2 - 2t) + 16 - 2(t^2 - 2t)^2 - 16$$

$$= -(t^2 - 2t)^2 + 8(t^2 - 2t)$$

$$= -(t^2 - 2t)(t^2 - 2t - 8)$$

$$= -t(t-2)(t+2)(t-4) > 0,$$

and we derive

$$t \in (-2, 0) \cup (2, 4). \qquad \textcircled{2}$$

Let x_1, x_2 be the x-coordinates of Q, R, respectively. By $\textcircled{1}$,

$$x_1 + x_2 = t^2 - 2t + 4, x_1 x_2 = \frac{1}{2}[(t^2 - 2t)^2 + 8].$$

Furthermore, since l is parallel to $y = x$,

$$\begin{aligned}
|PQ| \cdot |PR| &= \sqrt{2}(x_1 - t^2) \cdot \sqrt{2}(x_2 - t^2) \\
&= 2x_1 x_2 - 2t^2(x_1 + x_2) + 2t^4 \\
&= (t^2 - 2t)^2 + 8 - 2t^2(t^2 - 2t + 4) + 2t^4 \\
&= t^4 - 4t^3 + 4t^2 + 8 - 2t^4 + 4t^3 - 8t^2 + 2t^4 \\
&= t^4 - 4t^2 + 8 = (t^2 - 2)^2 + 4. \qquad \textcircled{3}
\end{aligned}$$

By $\textcircled{2}$, $t^2 - 2 \in (-2, 2) \cup (2, 14)$, $(t^2 - 2)^2 \in [0, 4) \cup (4, 196)$. So we conclude from $\textcircled{3}$ that

$$|PQ| \cdot |PR| = (t^2 - 2)^2 + 4 \in [4, 8) \cup (8, 200).$$

Remark 1   Alternatively, one can take the distance from the center of C_2 to line l, which must be smaller than the radius of C_2, to get $\left| \dfrac{4 + 2t - t^2}{\sqrt{2}} \right| < 2\sqrt{2}$, which also gives the range of t.

Remark 2   A shortcut of evaluating $|PQ| \cdot |PR|$ is to apply the secant theorem: since C_2 is centered at $M(4, 0)$, with radius $r = 2\sqrt{2}$, we have

$$|PQ| \cdot |PR| = |PM|^2 - r^2 = (t^2 - 4)^2 + (2t)^2 - (2\sqrt{2})^2 = t^4 - 4t^2 + 8. \quad \square$$

China Mathematical Competition (Complementary Test)

2016

---

1 (40 marks) Suppose real numbers $a_1, a_2, \ldots, a_{2016}$ satisfy $9a_i > 11a_{i+1}^2 (i = 1, 2, \ldots, 2015)$. Find the maximum of $(a_1 - a_2^2) \cdot (a_2 - a_3^2) \ldots (a_{2015} - a_{2016}^2) \cdot (a_{2016} - a_1^2)$.

Solution Let $P = (a_1 - a_2^2) \cdot (a_2 - a_3^2) \ldots (a_{2015} - a_{2016}^2) \cdot (a_{2016} - a_1^2)$. From the given condition,

$$a_i - a_{i+1}^2 > \frac{11}{9}a_{i+1}^2 - a_{i+1}^2 \geq 0 \ (i = 1, 2, \ldots, 2015).$$

When $a_{2016} - a_1^2 \leq 0$, we have $P \leq 0$.

Now assume $a_{2016} - a_1^2 > 0$. Define $a_{2017} = a_1$, then by the mean inequality we have

$$P^{\frac{1}{2016}} \leq \frac{1}{2016}\sum_{i=1}^{2016}(a_i - a_{i+1}^2) = \frac{1}{2016}\left(\sum_{i=1}^{2016}a_i - \sum_{i=1}^{2016}a_{i+1}^2\right)$$

$$= \frac{1}{2016}\left(\sum_{i=1}^{2016}a_i - \sum_{i=1}^{2016}a_i^2\right) = \frac{1}{2016}\sum_{i=1}^{2016}a_i(1 - a_i)$$

$$\le \frac{1}{2016} \sum_{i=1}^{2016} \left(\frac{a_i + (1 - a_i)}{2} \right)^2$$

$$= \frac{1}{2016} \cdot 2016 \cdot \frac{1}{4} = \frac{1}{4}.$$

Therefore, $P \le \dfrac{1}{4^{2016}}$.

When $a_1 = a_2 = \cdots = a_{2016} = \dfrac{1}{2}$, the equality holds, and

$$9a_i > 11a_{i+1}^2 (i = 1, 2, \ldots, 2015).$$

Therefore, the maximum of P is $\dfrac{1}{4^{2016}}$.   □

2 (40 marks) As shown in Fig. 2.1, in $\triangle ABC$, X and Y are two points on straight line BC (X, B, C, Y are arranged in order), so that

$$BX \cdot AC = CY \cdot AB.$$

Let the circumcenters of $\triangle ACX$ and $\triangle ABY$ be O_1 and O_2, respectively; and line O_1O_2 intersects with AB, AC at points U, V, respectively.

Prove: $\triangle AUV$ is an isosceles triangle.

Solution 1 As shown in Fig. 2.2, the internal angle bisector of $\angle BAC$ intersects BC at point P. Let the circumscribed circles of triangles ACX and ABY be ω_1 and ω_2, respectively. By the property of the internal angle bisector, $\dfrac{BP}{CP} = \dfrac{AB}{AC}$; and by the given condition, $\dfrac{BX}{CY} = \dfrac{AB}{AC}$. Then we have

$$\frac{PX}{PY} = \frac{BX + BP}{CY + CP} = \frac{AB}{AC} = \frac{BP}{CP}.$$

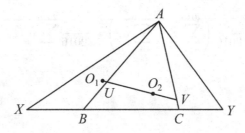

Fig. 2.1

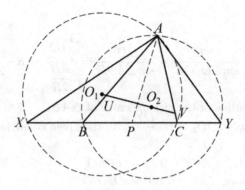

Fig. 2.2

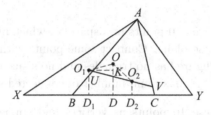

Fig. 2.3

Therefore, $CP \cdot PX = BP \cdot PY$. So P has the same power to ω_1 and ω_2, which means P is on the radical axis of ω_1 and ω_2, and then $AP \perp O_1O_2$. Therefore $\triangle AUV$ is an isosceles triangle. $\qquad \square$

Solution 2 As seen in Fig. 2.3, let the circumcenter of $\triangle ABC$ be O. Connect OO_1 and OO_2. Crossing O, O_1, O_2 draw perpendicular lines of BC, with foot points being D, D_1, D_2, respectively. Through O_1 draw $O_1K \perp OD$ at point K.

We will prove that $OO_1 = OO_2$. In right triangle OKO_1, we have

$$OO_1 = \frac{O_1K}{\sin \angle O_1OK}.$$

By the property of circumcenters, $OO_1 \perp AC$. Then $\angle O_1OK = \angle ACB$, as $OD \perp BC$. Since D, D_1 are the midpoints of BC, CX, respectively, we have

$$DD_1 = CD_1 - CD = \frac{1}{2}CX - \frac{1}{2}BC = \frac{1}{2}BX.$$

Therefore,

$$OO_1 = \frac{O_1K}{\sin \angle O_1OK} = \frac{DD_1}{\sin \angle ACB} = \frac{\frac{1}{2}BX}{\frac{AB}{2R}} = R \cdot \frac{BX}{AB},$$

where R is the radius of the circumcenter of $\triangle ABC$.

In the same way, $OO_2 = R \cdot \dfrac{CY}{AC}$. By the given conditioin $\dfrac{BX}{AB} = \dfrac{CY}{AC}$, we then get $OO_1 = OO_2$.

Since $OO_1 \perp AC$, $\angle AVU = 90° - \angle OO_1O_2$.

In the same way, $\angle AUV = 90° - \angle OO_2O_1$.

Furthermore, $\angle OO_1O_2 = \angle OO_2O_1$ as $OO_1 = OO_2$.

Therefore $\angle AUV = \angle AVU$, then $AU = AV$, and that means $\triangle AUV$ is an isosceles triangle. □

3 (50 marks) Given 10 points in a space, of which any four points are not on the same plane. Connect some points with line segments. If the obtained figure has no triangles and no space quadrilaterals, try to determine the maximum number of connected segments.

Solution Using these 10 points as vertices and connected segments as edges, we obtain a simple graph G of order 10. We prove that the number of edges of G does not exceed 15.

Let G's vertices be $v_1, v_2, \ldots, v_{10}$, have k edges in total, and use $\deg(v_i)$ to represent the degree of vertex v_i. If $\deg(v_i) \leq 3$ $(i = 1, 2, \ldots, 10)$ are true, then

$$k = \frac{1}{2} \sum_{i=1}^{10} \deg(v_i) \leq \frac{1}{2} \times 10 \times 3 = 15.$$

Otherwise, we may assume that $\deg(v_1) = n \geq 4$, and v_1 is adjacent to $v_2, \ldots, v_{n+1}$. Then there are no edges between $v_2, \ldots, v_{n+1}$, since no triangle is formed here. So there are exactly n edges among $v_1, v_2, \ldots, v_{n+1}$. For each j $(n+2 \leq j \leq 10)$, v_j is adjacent to at most one of $v_2, v_3, \ldots, v_{n+1}$ (otherwise, if v_j adjacent to v_s, v_t $(2 \leq s < t \leq n+1)$, then v_1, v_s, v_j, v_t constitute the four vertices of a spatial quadrilateral, which contradicts the conditions), So there are at most $10 - (n+1) = 9 - n$ edges between $v_2, \ldots, v_{n+1}$ and $v_{n+2}, \ldots, v_{10}$.

As there is no triangle among $v_{n+2}, \ldots, v_{10}$, then the number of edges there is at most $\left\lceil \dfrac{(9-n)^2}{4} \right\rceil$ according to Turán's theorem. Therefore, the

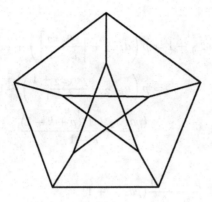

Fig. 3.1

number of edges of G

$$k \le n + (9 - n) + \left\lceil \frac{(9 - n)^2}{4} \right\rceil = 9 + \left\lceil \frac{(9 - n)^2}{4} \right\rceil$$

$$\le 9 + \left\lceil \frac{25}{4} \right\rceil = 15.$$

The graph shown in the figure has a total of 15 edges and meets the requirements.

In summary, the maximum number of edges of G is 15. □

4 (50 marks) It is know that p and $p + 2$ are prime numbers, $p > 3$, and sequence $\{a_n\}$ is defined as $a_1 = 2$, $a_n = a_{n-1} + \left\lceil \dfrac{pa_{n-1}}{n} \right\rceil$, $n = 2, 3, \ldots$, where $+\lceil x \rceil$ denotes the minimum integer not less than x. Prove: $n \mid pa_{n-1} + 1$, for any $n = 3, 4, \ldots, p - 1$.

Solution Notice that $\{a_n\}$ is an integer sequence. Use mathematical induction for n. When $n = 3$, $a_2 = 2 + p$ is known from the condition, so $pa_2 + 1 = (p+1)^2$. Since both p and $p + 2$ are prime numbers, and $p > 3$, we have $3 \mid p + 1$, so $3 \mid pa_2 + 1$; then the conclusion holds when $n = 3$.

When $3 < n \le p - 1$, and $k \mid pa_{k-1} + 1$ for $k = 3, \ldots, n - 1$, we have

$$\left\lceil \frac{pa_{k-1}}{k} \right\rceil = \frac{pa_{k-1} + 1}{k}.$$

Then

$$pa_{k-1} + 1 = p\left(a_{k-2} + \left\lceil \frac{pa_{k-2}}{k-1} \right\rceil\right) + 1$$

$$= p\left(a_{k-2} + \frac{pa_{k-2} + 1}{k-1}\right) + 1$$

$$= \frac{(pa_{k-2} + 1)(p + k - 1)}{k-1}.$$

Therefore,

$$pa_{n-1} + 1 = \frac{p+n-1}{n-1}(pa_{n-2} + 1)$$

$$= \frac{p+n-1}{n-1} \cdot \frac{p+n-2}{n-2}(pa_{n-3} + 1)$$

$$= \cdots = \frac{p+n-1}{n-1} \cdot \frac{p+n-2}{n-2} \cdots \frac{p+3}{3}(pa_2 + 1).$$

That means $pa_{n-1} + 1 = \dfrac{2n(p+1)}{(p+n)(p+2)} C_{p+n}^{n}$. From this (note that C_{p+n}^{n} is an integer),

$$n \mid (p+n)(p+2)(pa_{n-1} + 1). \qquad \text{①}$$

Since $n < p$ and p is a prime number,

$$(n, n+p) = (n, p) = 1.$$

Since $p+2$ is a prime number greater than n, $(n, p+2) = 1$. Therefore, n is prime with $(p+n)(p+2)$.

By ① we know then $n \mid pa_{n-1} + 1$.

The proof by mathematical induction is completed. $\qquad \square$

China Mathematical Competition (Complementary Test)

2017

Test Paper A, the Second Round

1 (40 marks) As shown in Fig. 1.1, in $\triangle ABC$, $AB = AC$, and I is the incenter. Let circle Γ_1 have center A and radius AB, and circle Γ_2 have center I and radius IB. A circle Γ_3 passing through B and I

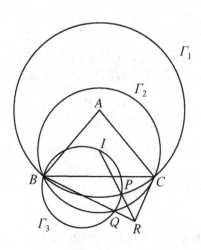

Fig. 1.1

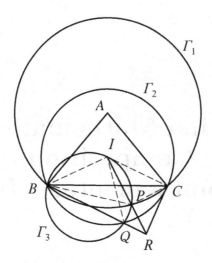

Fig. 1.2

intersects Γ_1, Γ_2 at P, Q, respectively (different from B). Let lines IP and BQ intersect at R.

Prove: $BR \perp CR$.

Solution Connect IB, IC, IQ, PB, and PC, as shown in Fig. 1.2.

Since Q is on Γ_2, $IB = IQ$, and

$$\angle IBQ = \angle IQB.$$

Since B, I, P, Q are concyclic, $\angle IQB = \angle IPB$. Thus $\angle IBQ = \angle IPB$. It follows that $IBP \backsim \triangle IRB$, $\angle IRB = \angle IBP$, and

$$\frac{IB}{IR} = \frac{IP}{IB}.$$

On the other hand, $AB = AC$, and I is the incenter of $\triangle ABC$. It follows $IB = IC$,

$$\frac{IC}{IR} = \frac{IP}{IC},$$

and this further implies that $\triangle ICP \backsim \triangle IRC$, $\angle IRC = \angle ICP$.

Notice that P lies on $\overparen{BC}$ of Γ_1, $\angle BPC = 180° - \dfrac{1}{2}\angle A$. We have

$$\angle BRC = \angle IRB + \angle IRC = \angle IBP + \angle ICP$$

$$= 360° - \angle BIC - \angle BPC$$

$$= 360° - \left(90° + \frac{1}{2}\angle A\right) - \left(180° - \frac{1}{2}\angle A\right)$$

$$= 90°.$$

Hence $BR \perp CR$. $\qquad\qquad\qquad\qquad\qquad\qquad\qquad\qquad\qquad\qquad$ □

2   Define sequence $\{a_n\}$ by $a_1 = 1$ and

$$a_{n+1} = \begin{cases} a_n + n, \ a_n \le n, \\ a_n - n, \ a_n > n, \end{cases} \quad n = 1, 2, \ldots.$$

Find the number of positive integer r's satisfying $a_r < r \le 3^{2017}$.

Solution   Clearly, $a_1 = 1, a_2 = 2$. Suppose for some $r \ge 2$, $a_r = r$. We shall prove by induction that for $t = 1, \ldots, r - 1$,

$$a_{r+2t-1} = 2r + t - 1 > r + 2t - 1, a_{r+2t} = r - t < r + 2t. \qquad ①$$

When $t = 1$, $a_r = r \ge r$, by definition:

$$a_{r+1} = a_r + r = r + r = 2r > r + 1,$$

$$a_{r+2} = a_{r+1} - (r+1) = 2r - (r+1) = r - 1 < r + 2,$$

and the conclusion follows.

Suppose ① holds for some $1 \le t < r - 1$. Then by definition:

$$a_{r+2t+1} = a_{r+2t} + (r + 2t) = r - t + r + 2t = 2r + t > r + 2t + 1,$$

$$a_{r+2t+2} = a_{r+2t+1} - (r + 2t + 1) = 2r + t - (r + 2t + 1)$$

$$= r - t - 1 < r + 2t + 2,$$

so ① also holds for $t + 1$. The induction is now complete and ① holds for all $t = 1, 2, \ldots, r - 1$; in particular when $t = r - 1$, $a_{3r-2} = 1$, and

$$a_{3r-1} = a_{3r-2} + (3r - 2) = 3r - 1.$$

Take all r's with $a_r = r$ and denote by $r_1, r_2, \ldots$. From the above recurrence formula, we have

$$r_1 = 1, r_2 = 2, r_{k+1} = 3r_k - 1, \ k = 2, 3, \ldots$$

and

$$r_{k+1} - \frac{1}{2} = 3\left(r_k - \frac{1}{2}\right) (k = 1, \ldots, m-1).$$

Thus

$$r_m = 3^{m-1}\left(r_1 - \frac{1}{2}\right) + \frac{1}{2} = \frac{3^{m-1} + 1}{2}.$$

From

$$r_{2018} = \frac{3^{2017} + 1}{2} < 3^{2017} < \frac{3^{2018} + 1}{2} = r_{2019},$$

we deduce that among indices $1, 2, \ldots, 3^{2017}$, there are 2018 r's with $a_r = r$, namely $r_1, r_2, \ldots, r_{2018}$.

From $①$, we see for every $k = 1, 2, \ldots, 2017$, exactly half of the indices $r_k + 1, r_k + 2, \ldots, 3r_k - 2$ satisfy $a_r < r$. Due to

$$r_{2018} + 1 = \frac{3^{2017} + 1}{2} + 1$$

and 3^{2017} both being odd, and all odd indices in $r_{2018} + 1, \ldots, 3^{2017}$ satisfy $a_r > r$, all even indices satisfy $a_r < r$, and furthermore there is 1 fewer even index than the odd indices, it follows that the number of positive r's that satisfy $a_r < r \le 3^{2017}$ is

$$\frac{1}{2}(3^{2017} - 2018 - 1) = \frac{3^{2017} - 2019}{2}. \qquad \square$$

③ (50 marks) Each unit square of a 33×33 square grid is colored by one of three colors such that the number of squares of each color is the same. A common side shared by two unit squares is called a "separating side" if the squares have different colors. Find the minimum number of separating sides.

Solution Let L be the number of separating sides. Divide the 33×33 square grid into three parts by the bold lines as shown in the figure, and color the three parts with the three colors, respectively. Then we have $L = 56$.

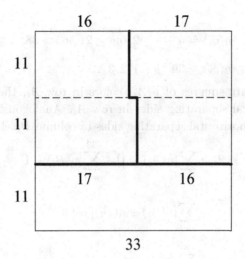

It is necessary to prove $L \geq 56$. Denote $A_1, A_2, \ldots, A_{33}$ as the rows of the grid from top to bottom, $B_1, B_2, \ldots, B_{33}$ as the columns from left to right. Let $n(A_i)$ be the number of colors appearing in row A_i, and $n(B_i)$ be the number of colors appearing in column B_i. Let c_1, c_2, c_3 be the three colors. For each color c_j, let $n(c_j)$ be the sum of the numbers of rows and columns that contain square(s) with color c_j. Define

$$\delta(A_i, c_j) = \begin{cases} 1, & \text{if } c_j \text{ appears in row } A_i, \\ 0, & \text{otherwise.} \end{cases}$$

Similarly, define $\delta(B_i, c_j)$ for each column and each color. Then

$$\sum_{i=1}^{33} [n(A_i) + n(B_i)] = \sum_{i=1}^{33} \sum_{j=1}^{3} [\delta(A_i, c_j) + \delta(B_i, c_j)]$$

$$= \sum_{j=1}^{3} \sum_{i=1}^{33} [\delta(A_i, c_j) + \delta(B_i, c_j)]$$

$$= \sum_{j=1}^{3} n(c_j).$$

There are $\frac{1}{3} \cdot 33^2 = 363$ squares of color c_j in the grid. If there are a rows and b columns in which c_j appears, then all squares of color c_j must be in the intersections of these rows and columns, which implies that $ab \geq 363$, and thereby

$$n(c_j) = a + b \geq 2\sqrt{ab} \geq 2\sqrt{363} > 38,$$

$$n(c_j) \geq 39, \ j = 1, 2, 3. \qquad \qquad ①$$

Since there are squares of $n(A_i)$ colors in row A_i, there are at least $n(A_i) - 1$ vertical separating sides in row A_i. Analogously, there are at least $n(B_j) - 1$ horizontal separating sides in column B_j. It follows that

$$L \geq \sum_{i=1}^{33} [n(A_i) - 1] + \sum_{i=1}^{33} [n(B_i) - 1]$$

$$= \sum_{i=1}^{33} [n(A_i) + n(B_i)] - 66 \qquad \qquad ②$$

$$= \sum_{j=1}^{3} n(c_j) - 66. \qquad \qquad ③$$

There are two cases.

Case 1: There exists a row (or a column) in which all the squares are of the same color, say c_1 for some row. Notice that all the 33 columns contain squares of color c_1, and at least 11 rows contain squares of color c_1 ($363 = 33 \times 11$). Thus

$$n(c_1) \geq 11 + 33 = 44. \qquad \qquad ④$$

By ①, ③ and ④, we get

$$L \geq n(c_1) + n(c_2) + n(c_3) - 66 \geq 44 + 39 + 39 - 66 = 56.$$

Case 2: There are no row or column that consists of the squares with the same color. Then for every $1 \leq i \leq 33$, $n(A_i) \geq 2$, $n(B_i) \geq 2$. By ②, we get

$$L \geq \sum_{i=1}^{33} [n(A_i) + n(B_i)] - 66 \geq 33 \times 4 - 66 = 66 > 56.$$

We conclude that the minimum number of separating sides is 56. □

4 (50 marks) Let $m \geq n \geq 1$ be integers, $a_1, a_2, \ldots, a_n$ be distinct positive integers not exceeding m, and $a_1, a_2, \ldots, a_n$ be coprime. Prove that for every real number x, there exists i ($1 \leq i \leq n$), such that $\|a_i x\| \geq \dfrac{2}{m(m+1)} \|x\|$, where $\|y\|$ denotes the distance from y to its nearest integer.

Solution First, we need two lemmas.

Lemma 1 *There exist integers $c_1, c_2, \ldots, c_n$, satisfying*

$$c_1 a_1 + c_2 a_2 + \cdots + c_n a_n = 1,$$

and $|c_i| \le m$, $1 \le i \le n$.

Proof of lemma 1 Since $(a_1, a_2, \ldots, a_n) = 1$, by Bézout's identity, there exist integers $c_1, c_2, \ldots, c_n$, satisfying

$$c_1 a_1 + c_2 a_2 + \cdots + c_n a_n = 1. \qquad ①$$

It suffices to modify certain values of $c_1, c_2, \ldots, c_n$ satisfying ①, so that no c_i's absolute value exceeds m. Define

$$S_1(c_1, c_2, \ldots, c_n) = \sum_{c_i > m} c_i \ge 0,$$

$$S_2(c_1, c_2, \ldots, c_n) = \sum_{c_j < -m} |c_j| \ge 0.$$

If $S_1 > 0$, there exists some $c_i > m > 1$, $c_i a_i > 1$. Since $a_1, a_2, \ldots, a_n$ are positive, from ① we have $c_j < 0$ for some j. Modify the values by taking

$$c_i' = c_i - a_j, c_j' = c_j + a_i, c_k' = c_k (1 \le k \le n, \ k \ne i, j).$$

We have

$$c_1' a_1 + c_2' a_2 + \cdots + c_n' a_n = 1, \qquad ②$$

and $0 \le m - a_j \le c_i' < c_i$, $c_j < c_j' < a_i \le m$. From $c_i' < c_i$, $c_j' < m$, it follows that

$$S_1(c_1', c_2', \ldots, c_n') < S_1(c_1, c_2, \ldots, c_n).$$

Further, from $c_j' > c_j$, $c_i' > 0$, it follows that

$$S_2(c_1', c_2', \ldots, c_n') \le S_2(c_1, c_2, \ldots, c_n).$$

If $S_2 > 0$, there exists some $c_j < -m$, and also some $c_i > 0$. Modify the values by taking

$$c_i' = c_i - a_j, c_j' = c_j + a_i, \ c_k' = c_k (1 \le k \le n, \ k \ne i, j).$$

We observe that ② holds, in which $-m < c_i' < c_i$, $c_j < c_j' < 0$. It is straightforward to check

$$S_1(c_1', c_2', \ldots, c_n') \leq S_1(c_1, c_2, \ldots, c_n),$$

and $S_2(c_1', c_2', \ldots, c_n') < S_2(c_1, c_2, \ldots, c_n)$.

Since S_1 and S_2 are nonnegative integers, after modifying the values finitely many times, we may obtain a set $c_1, c_2, \ldots, c_n$, satisfying ①, and $S_1 = S_2 = 0$. The lemma is proved.

Lemma 2 *(1) For any real numbers a, b, $\|a + b\| \leq \|a\| + \|b\|$.*

(2) For any integer u and real number y, $\|uy\| \leq |u| \cdot \|y\|$.

Proof of lemma 2 Observe that the equality $\|x + u\| = \|x\|$ holds for any real x and integer u. Let a, $b \in \left[-\dfrac{1}{2}, \dfrac{1}{2}\right]$, $\|a\| = |a|$, $\|b\| = |b|$. If $ab \leq 0$, assume that $a \leq 0 \leq b$, $a + b \in \left[-\dfrac{1}{2}, \dfrac{1}{2}\right]$ and we have

$$\|a + b\| = |a + b| \leq |a| + |b| = \|a\| + \|b\|.$$

If $ab > 0$, a, b are both positive, or both negative. If $|a| + |b| \leq \dfrac{1}{2}$, $a + b \in \left[-\dfrac{1}{2}, \dfrac{1}{2}\right]$ and we have

$$\|a + b\| = |a + b| = |a| + |b| = \|a\| + \|b\|.$$

If $|a| + |b| > \dfrac{1}{2}$, notice that $\|a + b\| \leq \dfrac{1}{2}$, and we have

$$\|a + b\| \leq \dfrac{1}{2} < |a| + |b| = \|a\| + \|b\|.$$

Therefore, (1) is verified. By (1) and $\|-y\| = \|y\|$, (2) is verified.

Now we return to the original problem. By lemma 1, there exist integers $c_1, c_2, \ldots, c_n$, satisfying

$$c_1 a_1 + c_2 a_2 + \cdots + c_n a_n = 1,$$

and $|c_i| \leq m$, $1 \leq i \leq n$. Hence

$$\sum_{i=1}^{n} c_i a_i x = x.$$

By lemma 2,

$$\|x\| = \left\| \sum_{i=1}^{n} c_i a_i x \right\| \leq \sum_{i=1}^{n} |c_i| \cdot \|a_i x\| \leq m \sum_{i=1}^{n} \|a_i x\|.$$

This implies that

$$\max_{1 \le i \le n} \|a_i x\| \ge \frac{1}{mn} \|x\|. \qquad (3)$$

If $n \le \frac{1}{2}(m+1)$, by ③ we have

$$\max_{1 \le i \le n} \|a_i x\| \ge \frac{\|x\|}{mn} \ge \frac{2\|x\|}{m(m+1)}.$$

If $n > \frac{1}{2}(m+1)$, there exist consecutive integers among $a_1, a_2, \ldots, a_n$, say a_1, a_2, satisfying

$$\|x\| = \|a_2 x - a_1 x\| \le \|a_2 x\| + \|a_1 x\|.$$

It follows that $\max\{\|a_2 x\|, \|a_1 x\|\} \ge \frac{\|x\|}{2} \ge \frac{2\|x\|}{m(m+1)}.$

Consequently, there always exists some $i(1 \le i \le n)$, satisfying $\|a_i x\| \ge \frac{2}{m(m+1)} \|x\|$. The proof is now completed. $\square$

Test Paper B, the Second Round

1 (40 marks) Suppose real numbers a, b, c satisfy $a + b + c = 0$. Let $d = \max\{|a|, |b|, |c|\}$. Prove

$$|(1+a)(1+b)(1+c)| \ge 1 - d^2.$$

Solution If $d \ge 1$, the inequality is obvious.

Suppose $0 \le d < 1$. Evidently, there are two of the given numbers that have the same sign, say a and b, and we have $ab \ge 0$. Then

$$(1+a)(1+b) = 1 + a + b + ab \ge 1 + a + b = 1 - c \ge 1 - d > 0,$$

and we deduce

$$|(1+a)(1+b)(1+c)| \ge |(1-c)(1+c)| = 1 - c^2 = 1 - |c|^2 \ge 1 - d^2. \quad \square$$

2 (40 marks) Given a positive integer m. Prove that $\mathbb{N}^+$ (the set of positive integers) can be divided into k disjoint subsets $A_1, A_2, \ldots, A_k$, such that for each $i = 1, \ldots, k$, A_i does not contain elements a, b, c, d (not necessarily distinct) that satisfy $ab - cd = m$.

Solution Take $k = m + 1$ and let $A_i = \{x \mid x \equiv i \pmod{m+1}, x \in \mathbb{N}^+\}$, $i = 1, 2, \ldots, m + 1$. For any $a, b, c, d \in A_i$, we have

$$ab - cd \equiv i \cdot i - i \cdot i = 0 \pmod{m+1}.$$

Therefore, $m + 1 \mid ab - cd$. But $m + 1$ does not divide m, hence $ab - cd \neq m$, completing the proof. □

3 (50 marks) As shown in the Fig. 3.1, in acute $\triangle ABC$, let ω be the circumcircle and D be the midpoint of major arc $\overparen{BC}$. Line DA and tangents of ω at B, C intersect at P, Q, respectively. Let BQ and AC intersect at X, CP and AB intersect at Y, BQ and CP intersect at T, as shown in Figure 1. Prove that AT bisects segment XY.

Solution First, we prove $YX // BC$, or $\dfrac{AX}{XC} = \dfrac{AY}{YB}$.
Connect BD, CD, as shown in Fig. 3.2. Since

$$\frac{S_{\triangle ACQ}}{S_{\triangle ABC}} \cdot \frac{S_{\triangle ABC}}{S_{\triangle ABP}} = \frac{S_{\triangle ACQ}}{S_{\triangle ABP}},$$

we have

$$\frac{\frac{1}{2} AC \cdot CQ \sin \angle ACQ}{\frac{1}{2} AB \cdot BC \sin \angle ABC} \cdot \frac{\frac{1}{2} AC \cdot BC \sin \angle ACB}{\frac{1}{2} AB \cdot BP \sin \angle ABP} = \frac{\frac{1}{2} AC \cdot AQ \sin \angle CAQ}{\frac{1}{2} AB \cdot AP \sin \angle BAP}.$$

$$\text{①}$$

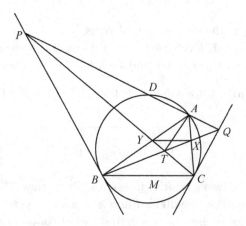

Fig. 3.1

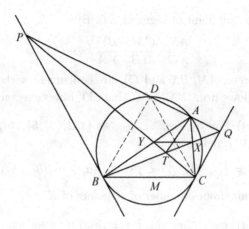

Fig. 3.2

As BP, CQ are tangents of ω, $\angle ACQ = \angle ABC$, $\angle ACB = \angle ABP$, and $\angle CAQ = \angle DBC = \angle DCB = \angle BAP$ (D is the midpoint of $\overset{\frown}{BC}$). It follows from ① that

$$\frac{AB \cdot AQ}{AC \cdot AP} = \frac{CQ}{BP}. \qquad ②$$

Since $\angle CAQ = \angle BAP$, we have $\angle BAQ = \angle CAP$, and

$$\frac{S_{\triangle ABQ}}{S_{\triangle ACP}} = \frac{\frac{1}{2} AB \cdot AQ \sin \angle BAQ}{\frac{1}{2} AC \cdot AP \sin \angle CAP} = \frac{AB \cdot AQ}{AC \cdot AP}, \qquad ③$$

$$\frac{S_{\triangle BCQ}}{S_{\triangle BCP}} = \frac{\frac{1}{2} BC \cdot CQ \sin \angle BCQ}{\frac{1}{2} BC \cdot BP \sin \angle CBP} = \frac{CQ}{BP}. \qquad ④$$

From ②, ③, ④, we get $\dfrac{S_{\triangle ABQ}}{S_{\triangle ACP}} = \dfrac{S_{\triangle CBQ}}{S_{\triangle BCP}}$, or

$$\frac{S_{\triangle ABQ}}{S_{\triangle CBQ}} = \frac{S_{\triangle ACP}}{S_{\triangle BCP}}.$$

Moreover, $\dfrac{S_{\triangle ABQ}}{S_{\triangle CBQ}} = \dfrac{AX}{XC}$, $\dfrac{S_{\triangle ACP}}{S_{\triangle BCP}} = \dfrac{AY}{YB}$, and thus $\dfrac{AX}{XC} = \dfrac{AY}{YB}$.

Let M be the midpoint of segment BC. By

$$\frac{AX}{XC} \cdot \frac{CM}{MB} \cdot \frac{BY}{YA} = 1$$

and Ceva's theorem, AM, BX and CY are concurrent with common intersection T. It follows from $YX // BC$ that AT bisects segment XY.   □

4 (50 marks) For $a_1, a_2, \ldots, a_{20} \in \{1, 2, \ldots, 5\}$, $b_1, b_2, \ldots, b_{20} \in \{1, 2, \ldots, 10\}$, define

$$X = \{(i, j) \mid 1 \le i < j \le 20,\ (a_i - a_j)(b_i - b_j) < 0\}.$$

Find the maximum number of elements of X.

Solution   Let $(a_1, a_2, \ldots, a_{20}, b_1, b_2, \ldots, b_{20})$ be a sequence such that $|X|$ is maximal.

For $k = 1, 2, \ldots, 5$, let t_k be the number of variables $a_1, \ldots, a_{20}$ taking value k. By the definition of X, if $a_i = a_j$, then $(i, j) \notin X$. So, at least $\sum_{k=1}^{5} C_{t_k}^2$ of (i, j)'s do not belong to X. Since $\sum_{k=1}^{5} t_k = 20$, by Cauchy inequality

$$\sum_{k=1}^{5} C_{t_k}^2 = \frac{1}{2} \cdot \left(\sum_{k=1}^{5} t_k^2 - \sum_{k=1}^{5} t_k \right) \ge \frac{1}{2} \cdot \left[\frac{1}{5} \left(\sum_{k=1}^{5} t_k \right)^2 - \sum_{k=1}^{5} t_k \right]$$

$$= \frac{1}{2} \cdot 20 \cdot \left(\frac{20}{5} - 1 \right) = 30.$$

Therefore, $|X|$ cannot exceed $C_{20}^2 - 30 = 190 - 30 = 160$.

On the other hand, take $a_{4k-3} = a_{4k-2} = a_{4k-1} = a_{4k} = k(k = 1, 2, \ldots, 5)$, $b_i = 6 - a_i$ $(i = 1, 2, \ldots, 20)$. For every i, j, $1 \le i < j \le 20$,

$$(a_i - a_j)(b_i - b_j) = (a_i - a_j)[(6 - a_i) - (6 - a_j)]$$

$$= -(a_i - a_j)^2 \le 0,$$

in which equality holds only when $a_i = a_j$, which happens exactly $5C_4^2 = 30$ times. In this instance, $|X| = C_{20}^2 - 30 = 160$.

So, max $|X| = 160$.   □

China Mathematical Olympiad

(Changsha, Hunan)

First Day
(8:00 – 12:30; November 23, 2016)

1 Define sequences $\{u_n\}$ and $\{v_n\}$ as follows:

$$u_0 = u_1 = 1, \quad u_n = 2u_{n-1} - 3u_{n-2} \quad \text{for} \quad n \geqslant 2;$$

$$v_0 = a, \ v_1 = b, \ v_2 = c, \ v_n = v_{n-1} - 3v_{n-2} + 27v_{n-3} \quad \text{for} \quad n \geqslant 3.$$

Assume that there exists a positive integer N, such that v_n is an integer divisible by u_n for all $n \geqslant N$. Prove that $3a = 2b + c$.

Proof Since $v_n = \dfrac{1}{27}(v_{n+3} - v_{n+2} + 3v_{n+1})$, it is easy to see by reverse induction that v_n are rational numbers for all $n \geq 0$, and in particular, a, b, c are rational. Using the characteristic equations, we may find the expression for u_n and v_n as follows:

$$u_n = \frac{1}{2}((1 + \sqrt{2}i)^n + (1 - \sqrt{2}i)^n),$$

$$v_n = c_1(-1 + 2\sqrt{2}i)^n + c_2(-1 - 2\sqrt{2}i)^n + c_3 3^n.$$

43

Setting $n = 0, 1, 2$ in v_n, we have

$$v_0 = c_1 + c_2 + c_3 = a, \qquad \text{①}$$

$$v_1 = (-1 + 2\sqrt{2}i)c_1 + (-1 - 2\sqrt{2}i)c_2 + 3c_3 = b, \qquad \text{②}$$

$$v_2 = (-7 - 4\sqrt{2}i)c_1 + (-7 + 4\sqrt{2}i)c_2 + 9c_3 = c. \qquad \text{③}$$

Solving this system of equations gives $c_1 = r + s\sqrt{2}i$, $c_2 = r - s\sqrt{2}i$, where r, s are rational numbers, and it following by ① that c_3 is also rational. Put $c_3 = t$. Thus

$$v_n = (r + s\sqrt{2}i)(-1 + 2\sqrt{2}i)^n + (r - s\sqrt{2}i)(-1 - 2\sqrt{2}i)^n + t3^n.$$

Let k be a positive integer such that kr, ks, kt are all integers. Define

$$w_n = 2k((r + s\sqrt{2}i)(1 + \sqrt{2}i)^n + (r - s\sqrt{2}i)(1 - \sqrt{2}i)^n).$$

Since $w_0 = 4kr \in \mathbb{Z}, w_1 = 4k(r - 2s) \in \mathbb{Z}$, and $\{w_n\}$ also satisfies the recursive relation $w_n = 2w_{n-1} - 3w_{n-2}$, hence $w_n \in \mathbb{Z}$ for all $n \geq 0$. We have

$$u_n w_n = k((r + s\sqrt{2}i)(-1 + 2\sqrt{2}i)^n + (r - s\sqrt{2}i)$$
$$\times (-1 - 2\sqrt{2}i)^n + 2r3^n) = kv_n + m3^n,$$

where $m = k(2r - t) \in \mathbb{Z}$. Hence $m3^n = u_n w_n - kv_n$. For $n \geq N$, $u_n \,|\, v_n$, it following that $u_n \,|\, m3^n$. Using induction, it is easy to see that $3 \nmid u_n$, thus $u_n \,|\, m$. If $m \neq 0$, the number of divisors of m is finite, hence $\{u_n\}$ is eventually periodic. However, $1 \pm \sqrt{2}i$ are not roots of unity, this can't happen. We conclude that $m = 0$, i.e., $kv_n = u_n w_n$. Thus $a = v_0 = \frac{1}{k}u_0 w_0 = 4r$, $b = v_1 = \frac{1}{k}u_1 w_1 = 4(r - 2s)$, $c = v_2 = \frac{1}{k}u_2 w_2 = 4(r + 4s)$. Therefore $3a = 2b + c = 12r$. This completes the proof. $\qquad \square$

2 As shown in the figure, in an acute triangle ABC with $AB > AC$, AY is the altitude from A, $\odot O$ and $\odot I$ are the circumcircle and incircle of triangle ABC respectively. $\odot I$ touches BC at point D, line AO meets BC at point X. The tangent lines of $\odot O$ at points B and C meet at point L. Let PQ be the diameter of $\odot O$ passing through I. Show that the points A, D, L are collinear if and only if the points P, X, Y, Q are concyclic.

Proof Denote the length of BC, CA, AB by a, b, c respectively, as put $p = \frac{1}{2}(a + b + c)$. It is well-known that $BD = p - b$, $CD = p - c$.

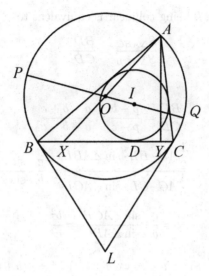

Fig. 2.1

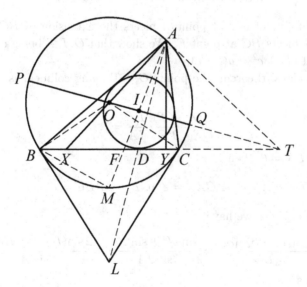

Fig. 2.2

The points A, D, L being collinear is equivalent to

$$\frac{S_{\triangle ABC}}{S_{\triangle ACL}} = \frac{BD}{CD}.$$

Noting that

$$\frac{BD}{CD} = \frac{p-b}{p-c} = \frac{a-b+c}{a+b-c},$$

$$\frac{S_{\triangle ABC}}{S_{\triangle ACL}} = \frac{\frac{1}{2} AB \cdot BL \cdot \sin \angle ABL}{\frac{1}{2} AC \cdot CL \cdot \sin \angle ACL} = \frac{c}{b} \cdot \frac{\sin \angle ABL}{\sin \angle ACL}$$

$$= \frac{c}{b} \cdot \frac{\sin \angle ACB}{\sin \angle ABC} = \frac{b^2}{c^2},$$

i.e., $\dfrac{c^2}{b^2} = \dfrac{a-b+c}{a+b-c}$. Since $b < c$, we get $b^2 + c^2 = a(b+c)$. Thus

$$A, D, L \text{ collinear} \Leftrightarrow b^2 + c^2 = a(b+c).$$

The tangent line of $\odot O$ at point A meets the extension of BC at point T. Line AI meets BC at point F. We show that O, I, T being collinear is equivalent to $b^2 + c^2 = a(b+c)$.

By Menelaus' theorem, the points O, I, T being collinear is equivalent to

$$\frac{AO}{OX} \cdot \frac{XT}{TF} \cdot \frac{FI}{IA} = 1. \tag{*}$$

Let $\angle ATX = \theta$, then

$$\theta = \angle ACB - \angle CAT = \angle ACB - \angle ABC = C - B.$$

By the law of sines, we have

$$\frac{AO}{OX} = \frac{S_{\triangle AOB} + S_{\triangle AOC}}{S_{\triangle BOC}} = \frac{\sin 2B + \sin 2C}{\sin 2A} = \frac{2 \sin(B+C) \cos(B-C)}{2 \sin A \cos A}$$

$$= \frac{\cos \theta}{\cos A}.$$

Let M be the midpoint of arc BC. Since $\triangle AFC \sim \triangle ABM$, and AT is tangent to $\odot O$, we have

$$\angle TAF = \angle ABM = \angle AFC,$$

thus $TF = TA$. It follows that $\dfrac{XT}{TF} = \dfrac{XT}{TA} = \dfrac{1}{\cos\theta}$. By the theorem of angle bisector, we have

$$\frac{FI}{IA} = \frac{BF}{c} = \frac{CF}{b} = \frac{BF+CF}{b+c} = \frac{a}{b+c}.$$

The left hand side of (*) is now

$$\frac{AO}{OX} \cdot \frac{XT}{TF} \cdot \frac{FI}{IA} = \frac{\cos\theta}{\cos A} \cdot \frac{1}{\cos\theta} \cdot \frac{a}{b+c} = \frac{1}{\cos A} \cdot \frac{a}{b+c}$$

$$= \frac{2abc}{(b+c)(b^2+c^2-a^2)}.$$

Noting that

$$(b+c)(b^2+c^2-a^2) - 2abc = (a+b+c)[a^2+b^2-c(a+b)],$$

thus (*) is true if and only if $b^2+c^2 = a(b+c)$. Therefore, A, D, L collinear is equivalent to O, I, T collinear.

If O, I, T is collinear, by the power of a point and the AY is the altitude of the right triangle AXT, we have

$$TQ \cdot TP = TA^2 = TY \cdot TX,$$

hence P, X, Y, Q are concyclic. Conversely, if P, X, Y, Q are concyclic, we extend PQ to meet line BC at point T'. It follows that

$$T'X \cdot T'Y = T'P \cdot T'Q = T'B \cdot T'C.$$

This equation uniquely determines T'. Since T also satisfies this equation, hence $T = T'$. This completes the proof. $\qquad\square$

3 A rectangle R is dissected into 2016 small rectangles with sides parallel to the sides of R. We call the vertices of those small rectangles *nodes*. A segment parallel to the sides of R is called basic if its two end points are nodes and there are no other nodes in the interior of it. Find the maximum and minimum of the number of basic segments among all possible dissections of R.

For example, in the figure on the right, R is dissected into 5 small rectangles with a total of 16 basic segments. The segments AB and BC are basic, while the segment AC is not.

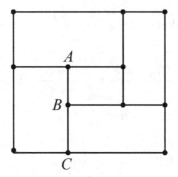

Solution We construct a simple graph G using the nodes as the vertices and the basic segments as the edges. Let N be the number of basic segments, which is the number of edges of G. It's easy to see that the degree of a vertex of G can only be 2, 3 or 4. There are exactly 4 vertices of degree 2, namely the four vertices of R. Let x and y denote the number of degree 3 and degree 4 vertices respectively. Summing up the degrees yields

$$N = \frac{1}{2}(4 \times 2 + 3x + 4y) = 4 + \frac{3}{2}x + 2y. \qquad ①$$

On the other hand, each small rectangle have 4 vertices. Counting with multiplicity, there are 2016×4 vertices. A degree 2 vertex belongs to exactly one small rectangle, a degree 3 vertex belongs to exactly two small rectangles, a degree 4 vertex belong to exactly 4 small rectangles.

Therefore,

$$2016 \times 4 = 4 + 2x + 4y,$$

i.e.,

$$x + 2y = 4030. \qquad ②$$

Plugging ② in ①, we have

$$N = 4034 + \frac{1}{2}x. \qquad ③$$

In view of ③, we see that maximizing (minimizing, resp.) N is equivalent to maximizing (minimizing, resp.) x, and is also equivalent to minimizing (maximizing, resp.) y by ②.

Consider the case that R is dissected by 2015 vertical segments. There is no degree 4 vertex, hence y achieves its minimum 0, and x achieve its maximum 4030. We conclude that $N_{\max} = 6049$.

Next we determine the minimum of N. Suppose that the basic segments inside R lies on s horizontal lines and t vertical lines. R is divided into at most $(s+1)(t+1)$ small rectangles, hence

$$(s+1)(t+1) \geq 2016.$$

Let ℓ be a vertical line which contains a basic segment e. Extend e as long as it remains on the boundary of some small rectangle until it reaches two endpoints, which must then be degree 3 vertices like $\top$ and $\perp$. We associated these two vertices to ℓ. Similarly each horizontal line that contains a basic segment is associated with two degree 3 vertices like $\vdash$ and $\dashv$. All these degree 3 vertices are distinct, hence

$$x \geq 2s + 2t.$$

Since

$$s + t = (s+1) + (t+1) - 2 \geq 2\sqrt{(s+1)(t+1)} - 2 \geq 2\sqrt{2016} - 2 > 87,$$

thus $s + t \geq 88$, $x \geq 176$. Plugging in ③ yields $N \geq 4122$.

$N = 4122$ may be achieved in the following example. First draw 44 horizontal lines and 44 vertical lines to divide R into $45^2 = 2025$ small rectangles. Then merge 10 consecutive small rectangles in the first row into one rectangle. This give an example of dissection of R into 2016 small rectangles with exactly 176 degree 3 vertices.

In conclusion, $N_{\max} = 6049$ and $N_{\min} = 4122$. $\qquad\square$

Second Day
(8:00 – 12:30; November 24, 2016)

4 Let $\alpha = (a_1, a_2, \ldots, a_n)$ and $\beta = (b_1, b_2, \ldots, b_n)$ be any two permutations of $1, 2, \ldots, n (n \geq 2)$. If a positive integer $k \leq n$ exists, such that

$$b_i = \begin{cases} a_{k+1-i}, & 1 \leq i \leq k, \\ a_i, & k+1 \leq i \leq n, \end{cases}$$

we then call α and β flip each other.

Prove: All permutations of $1, 2, \ldots, n$ can be properly recorded as $P_1, P_2, \ldots, P_m$, so that for each $i = 1, 2, \ldots, m$, P_i and P_{i+1} flip each other; here we stipulate $m = n!$ and $P_{m+1} = P_1$. (Contributed by WANG Xinmao)

Solution   Define mapping $f(1) = n$, $f(2) = 1$, $f(3) = 2, \ldots, f(n) = n - 1$. Let S be a set of all permutations of $1, 2, \ldots, n$. Define $1-1$ mapping of S to itself,

$$\phi : (a_1, a_2, \ldots, a_n) \mapsto (f(a_1), f(a_2), \ldots, f(a_n)).$$

It is easy to see that if α and β are flipped each other, then $\phi\,(\alpha)$ and $\phi\,(\beta)$ are also flipped each other.

We use mathematical induction to prove a slightly stronger conclusion: we can surround all m permutations of $1, 2, \ldots, n$ into a circle $P_1, P_2, \ldots, P_m$, such that

$$P_1 = (1, 2, \ldots, n - 1, n), P_m = (n, n - 1, \ldots, 2, 1),$$

and P_i, P_{i+1} flip each other, $1 \le i \le m - 1$.

When $n = 2$, the conclusion is obviously true. Suppose the conclusion is true for $n-1$ $(n > 2)$; that is, all $m_1 (=(n-1)!)$ permutations of $1, 2, \ldots, n-1$ can be arranged into a circle $Q_1, Q_2, \ldots, Q_{m_1}$ such that

$$Q_1 = (1, 2, \ldots, n - 1), Q_{m_1} = (n - 1, \ldots, 2, 1),$$

and Q_i, Q_{i+1} flip each other, $1 \le i \le m_1 - 1$. We construct permutations of $1, 2, \ldots, n$ in the following way:

$$P_{km_1+i} = \begin{cases} (Q_i, n), & k = 0, \\ \phi^k(P_i), & k \ge 1. \end{cases}$$

Here $\phi^k = \underbrace{\phi \circ \cdots \circ \phi}_{k}$ is a composite mapping, $0 \le k \le n - 1$, $1 \le i \le m_1$.

It is easy to see that P_{km_1+i} has the form $(*, \ldots, *, n - k)$, so $\{P_{km_1+i} \mid 0 \le k \le n - 1, \ 1 \le i \le m_1\}$ contains all permutations of $1, 2, \ldots, n$. Note

$$
\begin{aligned}
P_1 &= (1, 2, \ldots, n - 1, n), \ldots, \\
P_{m_1} &= (n - 1, \ldots, 2, 1, n), \\
P_{m_1+1} &= \phi(P_1) = (n, 1, \ldots, n - 2, n - 1), \ldots, \\
P_{2m_1} &= \phi(P_{m_1}) = (n - 2, \ldots, 1, n, n - 1),
\end{aligned}
$$

$$\ldots\ldots$$

$$
\begin{aligned}
P_{(n-1)m_1 + 1} &= \phi^{n-1}(P_1) = (2, 3, \ldots, n, 1), \ldots, \\
P_{nm_1} &= \phi^{n-1}(P_{m_1}) = (n, \ldots, 3, 2, 1).
\end{aligned}
$$

Then from the fact that P_{m_1} and P_{m_1+1} flip each other, we get P_{km_1} and P_{km_1+1} flip each other. Therefore, P_i and P_{i+1} flip each other, $1 \le i \le nm_1 - 1$.

The proof by mathematical induction is completed.   □

5 Given a positive integer n, define D_n as the set of all positive factors of n. Find all such n that D_n can be written as a union of two disjoint subsets A and G, satisfying: A and G both contain at least three elements, the elements in A can be arranged into an arithmetic sequence, and those in G can be arranged into an geometric sequence. (Contributed by Qu Zhenhua)

Solution We will prove by reduction to absurdity that such n does not exist. Assume there exists a positive integer n satisfying the given condition. Let

$$A = \{a_1 < a_2 < \cdots < a_k\}, \quad G = \{g_1 < g_2 < \cdots < g_l\},$$

where $k, l \geq 3$, $k + l = |D_n| = \tau(n)$.

Firstly, $g_l = n$. Otherwise we have $a_{k-1} \leq \dfrac{n}{2}$, then $a_{k-2} = 2a_{k-1} - a_k \leq 0$, contradicting the given condition. Moreover, the common ratio of G:
$$\lambda = \frac{g_l}{g_{l-1}} = \frac{n}{g_{l-1}}$$ is a positive integer.

Secondly, n contains at least 2 different prime factors. Because if $n = p^\alpha$ (p is a prime number), then

$$D_n = \{1, p, p^2, \ldots, p^\alpha\}.$$

For any $0 \leq u < v < w \leq \alpha$, we have

$$p^v - p^u = p^u(p^{v-u} - 1) \neq p^v(p^{w-v} - 1) = p^w - p^v.$$

Therefore p^u, p^v, p^w cannot constitute a arithmetic sequence, i.e., D_n contains no three numbers which constitute a arithmetic sequence.

Let p be a prime factor of the common ratio λ, and $n = p^\alpha m = g_l$, $p \nmid m$. Then

$$\alpha = (l - 1) \cdot v_p(\lambda) + v_p(g_1) \quad \text{and} \quad \tau(m) > 1.$$

Let $A = A_1 \bigcup A_2$, where

$$A_1 = \{x \in A \,|\, p \text{ is not a factor of } x\} \quad \text{and} \quad A_2 = \{x \in A \,|\, p$$
$$\text{is a factor of } x\}.$$

Then it easy to see that

$$|A_1| = \tau(m) - \delta \text{ and } |A_2| = \alpha \cdot \tau(m) - (l - \delta).$$

($\delta = 1$ or 0, according to whether $v_p(g_1) = 0$ or not.). Therefore,

$$|A_2| - |A_1| = \alpha \cdot \tau(m) - (l - \delta) - \tau(m) + \delta$$
$$= ((l - 1) \cdot v_p(\lambda) + v_p(g_1)) \cdot \tau(m) - \tau(m) - l + 2\delta$$
$$\geq (l - 1) \cdot \tau(m) - \tau(m) - l + (v_p(g_1) \cdot \tau(m) + 2\delta) \quad (\because v_p(\lambda) \geq 1).$$

It is easy to check that $v_p(g_1) \cdot \tau(m) + 2\delta \geq 2$. Then

$$|A_2| - |A_1| \geq (l - 2) \cdot (\tau(m) - 1).$$

If $\tau(m) > 2$ or $l > 3$ or $v_p(\lambda) > 1$ or $v_p(g_1) > 1$, then $|A_2| - |A_1| \geq 2$, and that means there are at least a pair of elements in A which both have p as a factor and are adjacent when all the elements in A are arranged as a arithmetic sequence. Therefore, the common difference of the sequence must has p as a factor, and this is true only when all the elements in A contain p as a factor. It is impossible.

If $\tau(m) = 2$ and $l = 3$ and $v_p(\lambda) = 1$, then m is a prime number. It is easy to see that either $A = \{1, p, p^2\}$ (case $v_p(g_1) = 0$) or $A = \{1, m, p, p^2, p^3\}$ (case $v_p(g_1) = 1$). In neither cases, the elements in A can be arranged into an arithmetic sequence.

The proof is completed. $\qquad\square$

6 Given integer $n \geq 2$, and positive numbers $a < b$, let $x_1, x_2, \ldots, x_n \in [a, b]$. Find the maximum value of

$$\frac{\dfrac{x_1^2}{x_2} + \dfrac{x_2^2}{x_3} + \cdots + \dfrac{x_{n-1}^2}{x_n} + \dfrac{x_n^2}{x_1}}{x_1 + x_2 + \cdots + x_{n-1} + x_n}.$$

(Contributed by Li Ting)

Solution 1

Lemma 1 Let $a \leq u \leq v \leq b$. Then $\dfrac{\dfrac{u^2}{v} + \dfrac{v^2}{u}}{u + v} \leq \dfrac{\dfrac{a^2}{b} + \dfrac{b^2}{a}}{a + b}$.

Proof of lemma 1   we have

$$\frac{\dfrac{u^2}{v} + \dfrac{v^2}{u}}{u+v} = \frac{u^2 - uv + v^2}{uv} = \frac{u}{v} + \frac{v}{u} - 1,$$

and

$$\frac{\dfrac{a^2}{b} + \dfrac{b^2}{a}}{a+b} = \frac{a}{b} + \frac{b}{a} - 1.$$

As $f(t) = t + \dfrac{1}{t}$ increases monotonically on $[1, +\infty)$ and $1 \leq \dfrac{v}{u} \leq \dfrac{b}{a}$, we have $f\left(\dfrac{v}{u}\right) \leq f\left(\dfrac{b}{a}\right)$. The proof is completed.

Lemma 2   Let $a \leq u \leq v \leq b$. Then $\dfrac{u^2}{v} + \dfrac{v^2}{u} - u - v \leq \dfrac{a^2}{b} + \dfrac{b^2}{a} - a - b$.

Proof of lemma 2   We have

$$\left(\frac{a^2}{v} + \frac{v^2}{a} - a - v\right) - \left(\frac{u^2}{v} + \frac{v^2}{u} - u - v\right) = \left(\frac{a^2}{v} - \frac{u^2}{v}\right) + \left(\frac{v^2}{a} - \frac{v^2}{u}\right)$$

$$+ (-a + u) = (u - a)\left(-\frac{u+a}{v} + \frac{v^2}{au} + 1\right)$$

$$= (u - a)\left(\frac{v - a}{v} + \frac{v^3 - au^2}{auv}\right) \geq 0.$$

Therefore,

$$\frac{u^2}{v} + \frac{v^2}{u} - u - v \leq \frac{a^2}{v} + \frac{v^2}{a} - a - v. \qquad \text{①}$$

On the other hand, we have

$$\left(\frac{a^2}{b} + \frac{b^2}{a} - a - b\right) - \left(\frac{a^2}{v} + \frac{v^2}{a} - a - v\right) = \left(\frac{a^2}{b} - \frac{a^2}{v}\right) + \left(\frac{b^2}{a} - \frac{v^2}{a}\right)$$

$$+ (-b + v) = (b - v)\left(-\frac{a^2}{bv} + \frac{b+v}{a} - 1\right)$$

$$= (b - v)\left(\frac{b^2v - a^3}{abv} + \frac{v - a}{a}\right) \geq 0.$$

Therefore,

$$\frac{a^2}{v} + \frac{v^2}{a} - a - v \leq \frac{a^2}{b} + \frac{b^2}{a} - a - b. \qquad \text{②}$$

By ① and ②, the proof is completed.

Proof of the original question Rearrange $x_1, x_2, \ldots, x_n$ into an order sequence $y_1 \leq y_2 \leq \cdots \leq y_n$. Since $\sum_{i=1}^{n} x_i = \sum_{i=1}^{n} y_i$, by the sequence inequality we have

$$\frac{\sum_{i=1}^{n} \dfrac{x_i^2}{x_{i+1}}}{\sum_{i=1}^{n} x_i} \leq \frac{\sum_{i=1}^{n} \dfrac{y_i^2}{y_{n+1-i}}}{\sum_{t=1}^{n} y_i}. \tag{3}$$

Define $m = \left[\dfrac{n}{2}\right]$, for $i = 1, 2, \ldots, m$, by lemma 1 we have

$$\left(\frac{y_i^2}{y_{n+1-i}} + \frac{y_{n+1-i}^2}{y_i}\right) \cdot (a + b) \leq \left(\frac{a^2}{b} + \frac{b^2}{a}\right) \cdot (y_i + y_{n+1-i}).$$

Sum it on index i,

$$(a + b) \sum_{i=1}^{m} \left(\frac{y_i^2}{y_{n+1-i}} + \frac{y_{n+1-i}^2}{y_i}\right) \leq \left(\frac{a^2}{b} + \frac{b^2}{a}\right) \sum_{i=1}^{m} (y_i + y_{n+1-i}). \tag{4}$$

We discuss the following two cases.

Case 1: $n = 2m$. by ④ we have

$$\frac{\sum_{i=1}^{n} \dfrac{y_i^2}{y_{n+1-i}}}{\sum_{n=1}^{n} y_i} \leq \frac{\dfrac{a^2}{b} + \dfrac{b^2}{a}}{a + b} = \frac{a^2 - ab + b^2}{ab}.$$

Then by ③,

$$\frac{\sum_{i=1}^{n} \dfrac{x_i^2}{x_{i+1}}}{\sum_{n=1}^{n} x_i} \leq \frac{a^2 - ab + b^2}{ab}.$$

The equality holds, when we assign $x_1, x_2, \ldots, x_n$ the values of $a, b, a, b, \ldots, a, b$, respectively. Therefore, the required maximum value is $\dfrac{a^2 - ab + b^2}{ab}$

Case 2: $2 : n = 2m + 1$. Let

$$U = \sum_{i=1}^{m} \left(\frac{y_i^2}{y_{n+1-i}} + \frac{y_{n+1-i}^2}{y_i} \right), \quad V = \sum_{i=1}^{m} (y_i + y_{n+1-i}).$$

By ④ then

$$(a + b)U \le \left(\frac{a^2}{b} + \frac{b^2}{a} \right) V.$$

By lemma 2 we have

$$U - V \le m \left(\frac{a^2}{b} + \frac{b^2}{a} - a - b \right).$$

Then

$$\left[m \left(\frac{a^2}{b} + \frac{b^2}{a} \right) + a \right] \cdot (V + y_{m+1}) - [m(a + b) + a] \cdot (U + y_{m+1})$$

$$= m \left[\left(\frac{a^2}{b} + \frac{b^2}{a} \right) V - (a + b)U \right] + m \left[\left(\frac{a^2}{b} + \frac{b^2}{a} \right) - (a + b) \right]$$

$$\times y_{m+1} - a(U - V)$$

$$\ge m \left[\left(\frac{a^2}{b} + \frac{b^2}{a} \right) - (a + b) \right] y_{m+1} - am \left[\left(\frac{a^2}{b} + \frac{b^2}{a} \right) - (a + b) \right]$$

$$= m \left[\left(\frac{a^2}{b} + \frac{b^2}{a} \right) - (a + b) \right] \cdot (y_{m+1} - a) \ge 0.$$

Therefore,

$$\frac{\sum\limits_{i=1}^{n} \frac{y_i^2}{y_{n+1-i}}}{\sum\limits_{i=1}^{n} y_i} \le \frac{m \left(\frac{a^2}{b} + \frac{b^2}{a} \right) + a}{(m + 1)a + mb}.$$

By ③ then,

$$\frac{\sum\limits_{i=1}^{n} \frac{x_i^2}{x_{i+1}}}{\sum\limits_{i=1}^{n} x_i} \le \frac{m(a^3 + b^3) + a^2 b}{ab[(m + 1)a + mb]}.$$

The equality holds, if we let $x_1, x_2, \ldots, x_n$ be $a, a, b, a, b, \ldots, a, b$, respectively. Therefore, the required maximum value is $\dfrac{m(a^3 + b^3) + a^2 b}{ab[(m + 1)a + mb]}$ $(m = \dfrac{n - 1}{2})$.

Solution 2 Let $x_0 = x_n$, $x_{n+1} = x_1$, $x_{n+2} = x_2$. For any $i = 1, \ldots, n$, fix $x_1, \ldots, x_{i-1}, x_{i+1}, \ldots, x_n$, and let

$$A = \frac{x_1^2}{x_2} + \cdots + \frac{x_{i-2}^2}{x_{i-1}} + \frac{x_{i+1}^2}{x_{i+2}} + \cdots + \frac{x_n^2}{x_1}, B = x_1 + \cdots + x_{i-1}$$
$$+ x_{i+1} + \cdots + x_n.$$

Define $f(x) = \dfrac{A + \dfrac{x_{i-1}^2}{x} + \dfrac{x^2}{x_{i+1}}}{B + x}$. Then

$$f'(x) = \frac{1}{(B+x)^2}\left[\left(-\frac{x_{i-1}^2}{x^2} + \frac{2x}{x_{i+1}}\right)(B+x) - \left(A + \frac{x_{i-1}^2}{x} + \frac{x^2}{x_{i+1}}\right)\right]$$

$$= \frac{1}{(B+x)^2}\left(-\frac{Bx_{i-1}^2}{x^2} - \frac{2x_{i-1}^2}{x} - A + \frac{2Bx}{x_{i+1}} + \frac{x^2}{x_{i+1}}\right),$$

$$f''(x) = \frac{1}{(B+x)^4}\left[\left(\frac{2Bx_{i-1}^2}{x^3} + \frac{2x_{i-1}^2}{x^2} + \frac{2B}{x_{i+1}} + \frac{2x}{x_{i+1}}\right)(B+x)^2\right.$$

$$\left. -\left(-\frac{Bx_{i-1}^2}{x^2} - \frac{2x_{i-1}^2}{x} - A + \frac{2Bx}{x_{i+1}} + \frac{x^2}{x_{i+1}}\right) \times 2(B+x)\right]$$

$$= \frac{1}{(B+x)^3}\left[\left(\frac{2Bx_{i-1}^2}{x^3} + \frac{2x_{i-1}^2}{x^2} + \frac{2B}{x_{i+1}} + \frac{2x}{x_{i+1}}\right)(B+x)\right.$$

$$\left. + 2\left(\frac{Bx_{i-1}^2}{x^2} + \frac{2x_{i-1}^2}{x^2} + A\right) - 2\left(\frac{2Bx}{x_{i+1}} + \frac{x^2}{x_{i+1}}\right)\right].$$

Note that

$$\left(\frac{2B}{x_{i+1}} + \frac{2x}{x_{i+1}}\right)(B+x) - 2\left(\frac{2Bx}{x_{i+1}} + \frac{x^2}{x_{i+1}}\right) = \frac{2B^2}{x_{i+1}} > 0.$$

Therefore $f''(x) > 0$, which means, for $x \in [a, b]$, $f(x)$ reaches the maximum at either $x = a$ or $x = b$.

As the result obtained above is true for any $i = 1, \ldots, n$, we then just need handle the problem for $x_1, \ldots, x_n \in \{a, b\}$ in the following.

According to Cauchy inequality, we have

$$\left(\frac{x_1^2}{x_2} + \cdots + \frac{x_n^2}{x_1}\right)(x_2 + x_3 + \cdots + x_n + x_1) \geq (x_1 + \cdots + x_n)^2.$$

Therefore, the equation in the original problem is not less than 1 at any time. If there exists x, such that

$$x_i = x_{i+1} = b, x_{i+2} = a,$$

we define

$$X_1 = \sum_{j \neq i} \frac{x_j^2}{x_{j+1}}, \quad Y_1 = \sum_{j \neq i+1} x_j.$$

Let $x_1' = x_1, x_2' = x_2, \ldots, x_i' = x_i = b, x_{i+1}' = a, x_{i+2}' = x_{i+2} = a, \ldots, x_n' = x_n$, and define

$$X_2 = \sum_{j \neq i} \frac{x_j'^2}{x_{j+1}'} = X_1, \quad Y_2 = \sum_{j \neq i+1} x_j' = Y_1.$$

Then $(x_1, \ldots, x_n)$ corresponds to $\dfrac{X_1 + b}{Y_1 + b} = 1 + \dfrac{X_1 - Y_1}{Y_1 + b}$ (by the discussion above, we know $X_1 + b \geq Y_1 + b$, i.e., $X_1 \geq Y_1$), and $(x_1', \ldots, x_n')$ corresponds to

$$\frac{X_2 + a}{Y_2 + a} = 1 + \frac{X_2 - Y_2}{Y_1 + a} \geq 1 + \frac{X_1 - Y_1}{Y_1 + b}.$$

Therefore, when taking the maximum value, sequence $x_1, \ldots, x_n$, $x_{n+1} = x_1$ cannot have two consecutive elements with value b.

Assume that in sequence $x_1, \ldots, x_n$ there are m "isolated" elements with value b, which are separated from each other by the rest elements (all with value a). Then $m \leq \left[\dfrac{n}{2}\right]$. The corresponding ratio is

$$\frac{m\left(\dfrac{b^2}{a} + \dfrac{a^2}{b}\right) + (n - 2m)a}{mb + (n - m)a} = \frac{na + m\left(\dfrac{b^2}{a} + \dfrac{a^2}{b} - 2a\right)}{na + m(b - a)}$$

$$= \frac{na + \dfrac{m}{ab}(b - a)(b^2 + ab - a^2)}{na + m(b - a)}$$

$$= \frac{b^2 + ab - a^2}{ab} + \frac{na}{na + m(b - a)}$$

$$\times \left(1 - \frac{b^2 + ab - a^2}{ab}\right)$$

$$= \frac{b^2 + ab - a^2}{ab} + \frac{na}{na + m(b - a)} \times \frac{a^2 - b^2}{ab}.$$

Since $a^2 - b^2 < 0$, the equation above reaches the maximum when m takes the greatest value, i.e., when $m = \left[\dfrac{n}{2}\right]$. Therefore,

(1) case $n = 2m$, the maximum value is $\dfrac{a^2 - ab + b^2}{ab}$;

(2) case $n = 2m + 1$, the maximum value is $\dfrac{m(a^3 + b^3) + a^2 b}{ab[(m + 1)a + mb]}$. $\qquad \square$

China Mathematical Olympiad

2017 China Mathematical Olympiad (also named the 33rd National Mathematics Winter Camp for Middle School Students), hosted by the Olympic Committee of the Chinese Mathematical Society and undertaken by the Zhejiang Provincial Mathematical Society and Hangzhou Xuejun Middle School, was held in Hangzhou from November 13 to 17, 2017. Participants in this competition were 386 middle school students from 35 representative teams from 31 provinces, cities, and autonomous regions in the China Mainland, Hong Kong Special Administrative Region, Macau Special Administrative Region, Singapore and Russia. 128 students won the first prize, 142 students won the second prize, and 86 students won the third prize.

60 students were selected from this competition to form the national training team.

The members of the main examination committee of this winter camp are:

Zhou Qing (East China Normal University);

Xiong Bin (East China Normal University);

Yao Yijun (Fudan University);

Qu Zhenhua (East China Normal University);

An Jinpeng (Peking University);

Ai Yinghua (Tsinghua University);

Wang Xinmao (University of Science and Technology of China);

Li Ting (Sichuan University);

Li Ming (Nankai University);

Fu Yunhao (Guangdong Second Normal University);

Wang Bin (Chinese Academy of Sciences Academy of Mathematics and Systems Science).

First Day
(8:00 – 12:30; November 15, 2017)

1 For a positive integer n, let A_n denote the set of all primes p satisfying the following property: *There exist positive integers a, b such that both $\dfrac{a+b}{p}$ and $\dfrac{a^n+b^n}{p^2}$ are integers and are coprime to p.* If A_n is finite (possibly empty), let $f(n)$ denote the number of elements in A_n.

(i) Prove that A_n is finite if and only if $n \neq 2$;

(ii) Let k, m be positive odd numbers, and d be their greatest common divisor. Prove that

$$f(d) \leqslant f(k) + f(m) - f(km) \leqslant 2f(d).$$

Proof   (i) It is clear that $A_1 = \varnothing$. If $n = 2$, then for every prime p, if we let $a = p$, $b = p^2$, then $\dfrac{a+b}{p} = 1 + p$ and $\dfrac{a^n+b^n}{p^2} = 1 + p^2$ are integers coprime to p. Thus A_2 consists of all primes, and hence is infinite.

Assume that $n \geqslant 3$. Let p be a prime such that there are positive integers a, b with $v_p(a+b) = 1$ and $v_p(a^n+b^n) = 2$, where $v_p(l)$ denotes the exponent of the maximal power of p that divides l. Suppose that $a+b = sp$, where $(s, p) = 1$. If one of a and b were divisible by p, then so were the other one, and it would follow that $v_p(a^n + b^n) \geq n > 2$, which is absurd. Thus both a and b are coprime to p.

(a) If n is even, then $0 \equiv a^n + b^n = a^n + (sp - a)^n \equiv 2a^n \pmod{p}$. But $(a^n, p) = 1$. So we must have $p = 2$. It follows that A_n is finite. (In fact, it can be shown that $A_n = \varnothing$.)

(b) If n is odd, then $0 \equiv a^n + (sp - a)^n \equiv nspa^{n-1} \pmod{p^2}$. But $(sa^{n-1}, p) = 1$. So we must have $v_p(n) \geqslant 1$. It again follows that A_n is finite.

(ii) Let us first show that if n is odd, then

$$A_n = \{p \text{ prime} \mid v_n(p) = 1\}.$$

This is obvious if $n = 1$. Assume that $n \geqslant 3$. Let $p \in A_n$. In view of case (b) above, we have $v_p(n) \geqslant 1$. It follows that if a, b satisfy $v_p(a+b) = 1$ and $v_p(a^n + b^n) = 2$, and if $a + b = sp$, then

$$a^n + b^n = a^n + (sp - a)^n \equiv -\frac{n(n-1)}{2} s^2 p^2 a^{n-2} + nspa^{n-1}$$

$$\equiv nspa^{n-1} \pmod{p^3}.$$

Thus, if $v_p(n) \geqslant 2$, it would follow that $v_p(a^n + b^n) \geqslant 3$, which is absurd. Hence $v_n(p) = 1$. Conversely, if $v_p(n) = 1$, it is easy to see that $a = 1$, $b = p - 1$ satisfy $v_p(a + b) = 1$ and $v_p(a^n + b^n) = 2$. Hence $p \in A_n$.

For a prime p and a positive odd number n, define

$$\chi_p(n) = \begin{cases} 1, & v_p(n) = 1, \\ 0, & v_p(n) \neq 1. \end{cases}$$

Then

$$f(n) = \sum_p \chi_p(n),$$

where the summation $\sum_p$ is taken over all primes p. Note that if k, m are odd, then so are km and $d = (k, m)$. Thus, it suffices to prove that

$$\sum_p \chi_p(d) \leqslant \sum_p \chi_p(k) + \sum_p \chi_p(m) - \sum_p \chi_p(km) \leqslant 2 \sum_p \chi_p(d). \quad \text{①}$$

For this, it suffices to show that

$$\chi_p(d) \leqslant \chi_p(k) + \chi_p(m) - \chi_p(km) \leqslant 2\chi_p(d) \quad \text{②}$$

for every prime p.

If $v_p(d) = 0$, then one of $v_p(k)$ and $v_p(m)$, say $v_p(k)$, is equal to 0. This implies that $v_p(km) = v_p(m)$. Thus we have $\chi_p(k) = \chi_p(d) = 0$, $\chi_p(km) = \chi_p(m)$, and ② follows.

If $v_p(d) = 1$, then $\min \{v_p(k), v_p(m)\} = 1$. It follows that $v_p(km) \geqslant 2$, and hence $\chi_p(km) = 0$. Assume that $v_p(k) = 1$. Then $\chi_p(k) + \chi_p(m) - \chi_p(km) = 1 + \chi_p(m) \in \{1, 2\}$. But $\chi_p(d) = 1$. So ② also holds.

If $v_p(d) \geqslant 2$, then $v_p(k), v_p(m), v_p(km) \geqslant 2$, and thus $\chi_p(d) = \chi_p(k) = \chi_p(m) = \chi_p(km) = 0$. Hence ② again follows. $\qquad \square$

Remark   To obtain the equality $A_n = \{p \text{ prime} \mid v_n(p) = 1\}$ for odd n, one can also use the following *lift the exponent theorem*: If p is a prime and a, b are integers coprime to p with $p \mid a + b$, then $v_p(a^n + b^n) = v_p(a + b) + v_p(n)$ for every positive odd number n.

② Let n and k be positive integers, and consider the set

$$T = \{(x, y, z) \mid x, y, z \in \mathbb{Z}, \quad 1 \le x, y, z \le n\}$$

of n^3 integer points in the 3-dimensional Euclidean space with Cartesian coordinates. Suppose that $(3n^2 - 3n + 1) + k$ points in T are colored red, such that: *If P and Q are two red points in T with PQ parallel to any coordinate axis, then all the integer points in the line segment PQ are colored red.* Prove that there are at least k distinct cubes of side length 1 whose vertices are all red points.

Proof   Let's introduce some terminologies. A red cube is a unit cube such that its vertices are all painted red; a red square is a unit square parallel to the Oxy plane such that its vertices are all painted red; a red interval is a unit interval parallel to the x-axis such that its vertices are all painted red. Let R be the set of all red points, let R_c be the set of all the red points in the plane $z = c$, let R_{bc} be the set of all the red points in the line $y = b, z = c$.

We will need the following

Lemma   *For each integer $1 \le c \le n$, the number of red squares in R_c is not less than $|R_c| - (2n - 1)$.*

Proof of the lemma   Suppose there are $S(c)$ red squares and I red intervals in R_c. We will count I in two ways. On one hand, for each $1 \le b \le n$, if $R_{bc} = \varnothing$, suppose that P is the point with the least x-coordinate among all the points in R_{bc}, and Q is the point with the largest x-coordinate, then R_{bc} is the set of all the integer points in the line segment PQ. The number of red intervals in R_{bc} is equal to $|R_{bc}| - 1$, then we have

$$I = \sum_{R_{bc} \ne \varnothing} (|R_{bc}| - 1) \ge \left(\sum_{R_{bc} \ne \varnothing} |R_{bc}| \right) - n = |R_c| - n. \qquad \text{①}$$

On the other hand, each red interval projects to a unit interval in the x-axis, we classify all the red intervals according to their projections. For each unit interval u in the x-axis, suppose that the number of red intervals in R_c with projection u is I_u. If $I_u \ge 1$, suppose that the minimal and maximal y-coordinate of these I_u red intervals are y_0 and y_1 respectively, then $I_u \leqslant y_1 - y_0 + 1$. By the rule of the painting, for each $y \in [y_0, y_1]$, the interval $u \times \{y\} \times \{c\}$ is a red interval. Thus, for each integer $y \in [y_0, y_1 - 1]$, the square $u \times \{y, y+1\} \times \{c\}$ is a red square. The number of such red squares is $(y_1 - y_0)$, at least $(I_u - 1)$. From this we conclude that

$$S(c) \geqslant \sum_{I_u \geqslant 1} (I_u - 1) \geqslant \left(\sum_{I_u \geqslant 1} I_u \right) - (n-1) = I - (n-1). \qquad ②$$

Combine ① and ②, we get that

$$S(c) \geqslant I - (n-1) \geqslant |R_c| - n - (n-1) = |R_c| - (2n-1),$$

this complete the proof of the lemma.

Back to the proof of the question. Suppose that there are C red cubes and S red squares in R. On the one hand, by the above lemma, we have

$$S = \sum_{c=1}^{n} S(c) \geqslant \sum_{c=1}^{n} (|R_c| - (2n-1)). \qquad ③$$

On the other hand, we can classify all the red squares according to their projections to the Oxy plane. For each unit square v in the Oxy plane, suppose that the number of red squares in R with projection v is S_v. If $S_v \geqslant 1$, suppose that the minimal and maximal z-coordinate of these S_v red squares are z_0 and z_1 respectively, then $S_v \leqslant z_1 - z_0 + 1$. By the rule of the painting, for each $z \in [z_0, z_1]$, the square $v \times \{z\}$ is a red square. Thus, for each integer $z \in [z_0, z_1 - 1]$, the cube $v \times \{z, z+1\}$ is a red cube. The number of such red cubes is $(z_1 - z_0)$, at least $(S_v - 1)$. From this we conclude that

$$C \geqslant \sum_{S_v \geqslant 1} (S_v - 1) \geqslant \left(\sum_{S_v \geqslant 1} S_v \right) - (n-1)^2 = S - (n-1)^2, \qquad ④$$

Combine ③ and ④, we get that

$$C \geqslant S - (n-1)^2 \geqslant \sum_{c=1}^{n} (|R_c| - (2n-1)) - (n-1)^2$$

$$= \left(\sum_{c=1}^{n} |R_c| \right) - (3n^2 - 3n + 1) = k.$$

Thus we have completed the proof. $\qquad\qquad\square$

Second Day
(8:00 – 12:30; November 16, 2017)

3   Let q be a positive integer which is not a cube. Prove that there exists a positive real number c such that

$$\{nq^{\frac{1}{3}}\} + \{nq^{\frac{2}{3}}\} \geqslant cn^{\frac{-1}{2}}$$

holds for every positive integer n, where $\{x\}$ denotes the fractional part of a real number x.

Proof   Let us prove that the constant $c = (13q^{\frac{4}{3}})^{-1}$ satisfies the requirement. We show this by contradiction. Assume that there exists a positive integer n such that

$$\{nq^{\frac{1}{3}}\} + \{nq^{\frac{2}{3}}\} < cn^{-\frac{1}{2}}.$$

Let

$$l = [nq^{\frac{2}{3}}], \quad m = [nq^{\frac{1}{3}}].$$

Firstly, we claim that there are integers r, s, t such that

$$r^2 + s^2 \neq 0, \quad |r| \leqslant \sqrt{n}, \quad |s| \leqslant \sqrt{n}, \quad rl + sm + tn = 0.$$

In fact, consider the set of pairs of integers

$$S = \{(u, v) \in \mathbb{Z}^2 : 0 \leqslant u, v \leqslant \sqrt{n}\}.$$

Then $|S| = ([\sqrt{n}] + 1)^2 > n$. It follows that there exist distinct pairs of integers (u_1, v_1) and (u_2, v_2) in S such that

$$u_1 l + v_1 m \equiv u_2 l + v_2.$$

Hence, the integers

$$r = u_1 - u_2, \quad s = v_1 - v_2, \quad t = -\frac{rl + sm}{n}$$

satisfy the required property. Now, let us consider the function

$$F(x) = r^3 x^6 + s^3 x^3 + t^3 - 3rstx^3.$$

Then $F(q^{\frac{1}{3}})$ is an integer. Let

$$f(x) = rx^2 + sx + t, \quad \omega = e^{2\pi i/3}.$$

It is easy to verify that

$$F(x) = f(x)f(\omega x)f(\omega^2 x).$$

Since q is not a cube, each of the numbers $q^{\frac{1}{3}}, \omega q^{\frac{1}{3}}, \omega^2 q^{\frac{1}{3}}$ is not a root of the equation $f(x) = 0$, and hence $F(q^{\frac{1}{3}}) = 0$. It follows that $|F(q^{\frac{1}{3}})| \geqslant 1$. On the other hand, we have

$$|f(q^{\frac{1}{3}})| = |rq^{\frac{2}{3}} + sq^{\frac{1}{3}} + t| = \frac{1}{n}|r(nq^{\frac{2}{3}} - l) + s(nq^{\frac{1}{3}} - m)|$$

$$\leqslant \frac{1}{n}(|r|\{nq^{\frac{2}{3}}\} + |s|\{nq^{\frac{1}{3}}\}) < \frac{c}{n} = \frac{1}{13q^{\frac{4}{3}}n},$$

$$f(\omega q^{\frac{1}{3}})f(\omega^2 q^{\frac{1}{3}}) = |f(\omega q^{\frac{1}{3}})|^2 \leqslant (|f(\omega q^{\frac{1}{3}}) - f(q^{\frac{1}{3}})| + |f(q^{\frac{1}{3}})|)^2$$

$$\leqslant (|r(\omega^2 - 1)q^{\frac{2}{3}}| + |s(\omega - 1)q^{\frac{1}{3}}| + (13q^{\frac{4}{3}}n)^{-1})^2$$

$$\leqslant (\sqrt{n} \cdot \sqrt{3} \cdot q^{\frac{2}{3}} + \sqrt{n} \cdot \sqrt{3} \cdot q^{\frac{2}{3}}$$

$$+ \sqrt{n} \cdot q^{\frac{2}{3}}/13)^2$$

$$< 13q^{\frac{4}{3}}n.$$

This implies that $|F(q^{\frac{1}{3}})| < 1$, a contradiction. □

4 As shown in the Figure, $ABCD$ is a concyclic quadrilateral whose diagonals meet at point P. The circumcircle of triangle APD meets the segment AB at points A and E. The circumcircle of triangle BPC meets the segment AB at points B and F. Let I and J be the incenter of triangle ADE and triangle BCF, respectively. Segments IJ and AC meet at point K. Show that points A, I, K, E are concyclic.

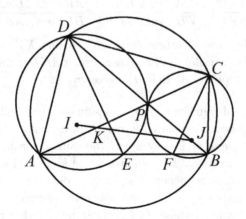

Fig. 4.1

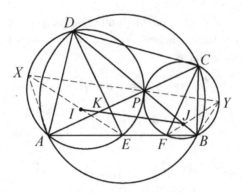

Fig. 4.2

Proof　Let the ray EI meet the circumcircle of $\triangle APD$ at point X, the ray FJ meet the circumcircle of $\triangle BPC$ at point Y. Draw segments PX, PY, XA and YB. By the well-known property of incenter, we have $XI = XA, YJ = YB$.

Clearly PX and PY are bisectors of $\angle APD$ and $\angle BPC$ respectively, thus X, P, Y are collinear. Since

$$\angle APX = \angle CPY = \angle BPY,$$

$$\angle AXP = \angle ADP = \angle BCP = \angle BYP,$$

triangle APX and BPY are similar, hence $XA/YB = AP/BP$. Thus

$$\frac{XI \sin \angle IXP}{YJ \sin \angle JYP} = \frac{XA}{YB} \cdot \frac{\sin \angle EAP}{\sin \angle FBP} = \frac{AP}{BP} \cdot \frac{\sin \angle BAP}{\sin \angle ABP} = 1,$$

which implies that the distances from I, J to line XY are equal. As a consequence, IJ is parallel to XY. Therefore, $\angle AKI = \angle APX = \angle AEI$, and it follows that points A, I, K, E are concyclic.　□

5　Given an odd number $n \geqslant 3$, and suppose that each square in the $n \times n$ chessboard is colored either black or white. Two squares are said to be *adjacent* if they are of the same color and have at least one common vertex. Two squares a and b are said to be *connected*, if there exists a sequence of squares $c_1, c_2, \ldots, c_k$ with $c_1 = a$ and $c_k = b$ such that c_i and c_{i+1} are adjacent for every $i = 1, \ldots, k-1$. Find the maximal number M for which there exists a coloring scheme admitting M pairwisely disconnected squares.

Solution   The answer is $M = \frac{1}{4}(n+1)^2 + 1$.

Consider the generalized problem on $m \times n$ chessboard, where $m, n \geqslant 3$ are both odd numbers. Suppose all the mn squares can be divided into K connected components, such that any two squares are connected if and only if they belong to the same connected component. We will prove by induction on (m, n) that (i) $K \leqslant \frac{1}{4}(m+1)(n+1) + 1$; (ii) If $K = \frac{1}{4}(m+1)(n+1) + 1$ then each squre at the four corners of the chessboard is connected to none of the other squares.

When $m = n = 3$, the 8 outside squares belongs to at most 4 connected components. Hence $K \leqslant 5$, and $K = 5$ if and only if the 4 squres at the corner are in the same color, and the other 5 squares are in another color.

Next assume $m \geqslant 5$. Suppose the second row of the chessboard can be divided into k parts. Each part consist of continuous squares in the same color. Let x_i be the number of squares in the ith part, $1 \leqslant i \leqslant k$. Let P be the number of connected components, which contain at least one square in the 1st row, but no square in the 2nd row. If $k \geqslant 2$ then we have $P \leqslant \left\lceil \frac{x_1 - 1}{2} \right\rceil + \left\lceil \frac{x_k - 1}{2} \right\rceil + \sum_{i=2}^{k} \left\lceil \frac{x_i - 2}{2} \right\rceil \leqslant \frac{n - k + 2}{2}$; If $k = 1$, we also have $P \leqslant \left\lceil \frac{n}{2} \right\rceil = \frac{n - k + 2}{2}$. Let Q be the number of connected components, which contain at least one square in the 2nd row, but no square in the 3rd row. We have $Q \leqslant \left\lceil \frac{k}{2} \right\rceil \leqslant \frac{k+1}{2}$. Let R be the number of connected components, which contain at least one square in the 3rd to n^{th} rows. By induction hypothesis (i), we have $R \leqslant \frac{1}{4}(m-1)(n+1) + 1$.

If $Q = \frac{k+1}{2} = 1$ then all the squares in the 2nd row are in the same color, all the squares in the 3rd row are also in the same color. If $Q = \frac{k+1}{2} \geqslant 2$, then the first square in the 3rd row is connected to the last square in the 3rd row via the $2i^{th}$ part of the 2nd row. By induction hypothesis (ii), $R \leqslant \frac{1}{4}(m-1)(n+1)$. Hence $Q = \frac{k+1}{2}$ and $R = \frac{1}{4}(m-1)(n+1) + 1$ can't holds simultaneously. Therefore, we have

$$K = P + Q + R \leqslant \frac{n - k + 2}{2} + \frac{k+1}{2} + \frac{1}{4}(m-1)(n+1)$$

$$= \frac{1}{4}(m+1)(n+1) + 1. \tag{$*$}$$

If the equality in $(*)$ holds, then we have $P = \dfrac{n-k+2}{2}$. Thus, the square at the upper-left corner is surrounded by three squres in opposite color. By symmetry, each squre at the four corners of the chessboard is connected to none of the other squares. It's easy to verify that the following coloring theme yields $K = \dfrac{1}{4}(m+1)(n+1) + 1$. The square at position (i,j) is colored black/white if ij is even/odd, respectively. □

> **6** Given positive integers n and k with $n > k$, and real numbers $a_1, a_2, \ldots, a_n \in (k-1, k)$. Suppose $x_1, x_2, \ldots, x_n$ are positive real numbers, satisfying $\sum_{i \in I} x_i \leqslant \sum_{i \in I} a_i$ for any k-element set $I \subset \{1, 2, \ldots, n\}$. Find the maximum of $x_1 x_2 \cdots x_n$.

Solution We show that the maximum is $a_1 a_2 \cdots a_n$. For $x_i = a_i$, $1 \leqslant i \leqslant n$, $x_1, x_2, \ldots, x_n$ satisfy the condition in the problem, and $x_1 x_2 \cdots x_n = a_1 a_2 \cdots a_n$. It remains to prove that $x_1 x_2 \cdots x_n \leqslant a_1 a_2 \cdots a_n$ in any case.

If $k = 1$, the result is trivial since $x_i \leqslant a_i$ for $1 \leqslant i \leqslant n$. In what follows we assume that $k \geqslant 2$.

Without loss of generality, we may assume $a_1 - x_1 \leqslant a_2 - x_2 \leqslant \cdots \leqslant a_n - x_n$. If $a_1 - x_1 \geqslant 0$, then $a_i \geqslant x_i$ for $1 \leqslant i \leqslant n$, and the result follows immediately. Now we assume $a_1 - x_1 < 0$.

Take $I = \{1, 2, \ldots, k\}$, by assumption we have

$$\sum_{i=1}^{k}(a_i - x_i) \geqslant 0, \tag{1}$$

hence $a_k - x_k \geqslant 0$. Thus, there exists s, $1 \leq s \leq k$, such that

$$a_1 - x_1 \leqslant \cdots \leqslant a_s - x_s < 0 \leqslant a_{s+1} - x_{s+1} \leqslant \cdots \leqslant a_k - x_k \leqslant \cdots \leqslant a_n - x_n.$$

Write $d_i = |a_i - x_i|$, $1 \leqslant i \leqslant n$, then

$$d_1 \geqslant \cdots \geqslant d_s \geqslant 0, 0 \leqslant d_{s+1} \leqslant \cdots \leqslant d_k \leqslant \cdots \leqslant d_n.$$

It follows from ① that $-d_1 - \cdots - d_s + d_{s+1} + \cdots + d_k \geqslant 0$, hence $d_{s+1} + \cdots + d_k \geqslant d_1 + \cdots + d_s$. Let $M = d_1 + \cdots + d_s$ and $N = d_{s+1} + \cdots + d_n$.

Note that $d_j < a_j < k$ for $j > s$. By GM-AM inequality,

$$\prod_{i=1}^{n} \frac{x_i}{a_i} = \prod_{i=1}^{s} \left(1 + \frac{d_i}{a_i}\right) \prod_{j=s+1}^{n} \left(1 - \frac{d_j}{a_j}\right)$$

$$\leqslant \prod_{i=1}^{s} \left(1 + \frac{d_i}{k-1}\right) \prod_{j=s+1}^{n} \left(1 - \frac{d_j}{k}\right)$$

$$\leqslant \left(1 + \frac{M}{s(k-1)}\right)^{s} \left(1 - \frac{N}{(n-s)k}\right)^{n-s}.$$

Since

$$\frac{N}{n-s} = \frac{d_{s+1} + \cdots + d_n}{n-s} \geq \frac{d_{s+1} + \cdots + d_k}{k-s} \geq \frac{M}{k-s},$$

we have

$$\prod_{i=1}^{n} \frac{x_i}{a_i} \leq \left(1 + \frac{M}{s(k-1)}\right)^{s} \left(1 - \frac{M}{(k-s)k}\right)^{n-s}$$

$$\leq \left(1 + \frac{M}{s(k-1)}\right)^{s} \left(1 - \frac{M}{(k-s)k}\right)^{k+1-s}.$$

We shall use the following two well-known inequalities

$$(1+x)^{\frac{1}{x}} < e, \quad (x > 0),$$

$$(1-x)^{\frac{1}{x}} < \frac{1}{e}, \quad (0 < x < 1).$$

Note that $M > 0$, it follows that

$$\left(1 + \frac{M}{s(k-1)}\right)^{2} < \exp\left(\frac{M}{k-1}\right),$$

$$\left(1 - \frac{M}{(k-s)k}\right)^{n-s} < \exp\left(-\frac{M(k+1-s)}{(k-s)k}\right).$$

Therefore

$$\prod_{i=1}^{n} \frac{x_i}{a_i} < \exp\left(\frac{M}{k-1} - \frac{M(k+1-s)}{(k-s)k}\right)$$

$$\leqslant \exp\left(\frac{M}{k-1} - \frac{M(k+1-s)}{(k+1-s)(k-1)}\right) = 1,$$

which completes the proof. $\qquad\square$

China National Team Selection Test

2017

1 Given integer $n \geq 3$, for a real number sequence $a_1, a_2, \ldots, a_n,$, if $1 \leq i < j < k \leq n$ satisfies $i+k = 2j$ and $a_i + a_k \neq 2a_j$, we say (i, j, k) forms a "non-arithmetic triple". For all real number sequences with n terms and containing non-arithmetic triple, find one that contains the minimum number of non-arithmetic triples. (Contributed by Wu Hao)

Solution Let $a_1 = 1, a_2 = a_3 = \cdots = a_n = 0$. Then only $(1, 1+d, 1+2d)$ $(0 < 2d \leq n-1)$ are non-arithmetic triples, whose number is $\left[\dfrac{n-1}{2}\right]$. We will prove by mathematical induction that $\left[\dfrac{n-1}{2}\right]$ is the required minimum number.

When $n = 3, 4$, the conclusion is obviously true. Assume that the conclusion is true for $n - 1 \geq 4$. Consider the case of n. Since $a_1, a_2, \ldots, a_n$ are not an arithmetic sequence, so either the first $n - 1$ terms or the last $n - 1$ terms are not an arithmetic sequence. We may assume that the latter is true. By induction, there are at least $\left[\dfrac{n-2}{2}\right]$ non-arithmetic triples (i, j, k), satisfying $2 \leq i < j < k \leq n$. The discussion is then divided into two subcases below.

(1) $a_3, a_4, \ldots, a_n$ are an arithmetic sequence. We may assume that $a_2 \neq 0$, $a_3 = a_4 = \cdots = a_n = 0$. If $a_1 = 0$, then a_1, a_2, a_3 are not arithmetic sequence and $(1, 2, 3)$ is a non-aithmetic triple; if $a_1 \neq 0$, then a_1, a_3, a_5

are not an arithmetic sequence and $(1, 3, 5)$ is a non-aithmetic triple. So the number of non-arithmetic triples is not less than $\left[\dfrac{n-2}{2}\right] + 1 = \left[\dfrac{n}{2}\right]$.

(2) $a_3, a_4, \ldots, a_n$ are not an arithmetic sequence, and by induction there are at least $\left[\dfrac{n-3}{2}\right]$ non-arithmetic triples (i, j, k) satisfying $3 \leq i < j < k \leq n$. We will prove later that there is at least one non-arithmetic triple (i, j, k) with i being either 1 or 2, so that the number of non-arithmetic triples in $a_1, a_2, \ldots, a_n$ is not less than $\left[\dfrac{n-3}{2}\right] + 1 = \left[\dfrac{n-1}{2}\right]$, confirming the conclusion.

Now we prove: if there are no non-arithmetic triples (i, j, k) with i being either 1 or 2 in sequence $a_1, a_2, \ldots, a_n$, then it is an arithmetic sequence. By mathematical induction, for sequence a_1, a_2, a_3, it is certainly true. Assume it is true for $a_1, a_2, \ldots, a_k (3 \leq k < n)$. If k is an odd number, then $a_2, a_{\frac{k+3}{2}}, a_{k+1}$ is an arithmetic sequence; if k is an even number, then $a_1, a_{\frac{k+2}{2}}, a_{k+1}$ is an arithmetic sequence. So in either case above, $a_1, a_2, \ldots, a_{k+1}$ is an arithemetic sequence. Therefore, $a_1, a_2, \ldots, a_n$. is an arithmetic sequence if it contains no non-arithmetic triples (i, j, k) with i being either 1 or 2.

The proof is completed. $\qquad\square$

2 Let $n = 2017$. Find the smallest positive integer m with the following properties: For any real coefficient polynomial $f(x)$, there are a real coefficient polynomial $g(x)$ with degree not greater than m, and n real numbers $a_1, a_2, \ldots, a_n$ different from each other, so that for $i = 1, 2, \ldots, n$,

$$g(a_i) = f(a_{i+1}),$$

where $a_{n+1} = a_1$. (Contributed by Wang Xinmao)

Solution For any function $\phi(x)$, define $\phi^n(x) = \underbrace{\phi(\cdots(\phi(x))\cdots)}_{n}$

First of all, we assert that $m \geq 2$. Otherwise, for $f(x) = x$, assume there are a polynomial of the first degree $g(x) = px + g$ and real numbers $a_1, a_2, \ldots, a_n$ satisfying the given condition. We have, for $i = 1, 2, \ldots, n, g(a_i) = f(a_{i+1}) = a_{i+1}$, and then $a_1 = f(a_{n+1}) = g^n(a_1)$.

(1) If $p \neq 1$, let $x_0 = -\dfrac{q}{p-1}$ be the fixed point of $g(x)$. Then from

$$a_1 = g^n(a_1) = p^n(a_1 - x_0) + x_0,$$

we have $a_1 = x_0$ and, by recursion, $a_i = g(x_0) = a_1$, for $i = 2, \ldots, n$, contradicting the assumption that $a_1, a_2, \ldots, a_n$ are different from each other.

(2) If $p = 1$, from $a_1 = g^n(a_1) = a_1 + nq$, we get $q = 0$ and $g(x) = x = f(x)$. Then, for $i = 1, 2, \ldots, n$, $a_i = g(a_i) = f(a_{i+1}) = a_{i+1}$. It is also impossible.

Now we prove that $m = 2$ can meet the given condtion. Let $G(x) = 4x(1-x)$. For any real coefficient polynomial $f(x)$ of degree l, excluding easy and similar cases, we may assume that $l \geq 1$, and there exists a constant $c > 0$ saisfying: $f(x) > 0, f'(x) > 0$, and $f''(x) \geq 0$ for $x > c$. Let $F(x) = \dfrac{f(x+c) - f(c)}{f(1+c) - f(c)}$. Then $F(x)$ is strickly monotone increasing on $[0, 1]$ satisfying $F(0) = 0, F(1) = 1$. Furthermore, $F'(x) > 0$ and $F''(x) > 0$, i.e., $F(x)$ is a concave function. So we have $\forall x \in \left[0, \dfrac{1}{2}\right]$, $F(x) \leq x < G(x)$.

Define $H(x) = F^{-1}(G(x))$. Then

(1) $H = \left[0, \dfrac{1}{2}\right] \to [0, 1]$ is a strickly monotone increasing bijective function;

(2) $\forall x \in \left(0, \dfrac{1}{2}\right)$, $H(x) > x$.

Let $b_0 = 1, b_1 = \dfrac{1}{2}, b_k = H^{-1}(b_{k-1})$ for $k \geq 2$. We have

$$0 < b_n < b_{n-1} < \cdots < b_1 < b_0 = 1, \text{ and } H^n(b_n) = b_0 = 1.$$

Let $L(x) = H^n(x) - 1 + x$. Then $L(0) = -1 < 0, L(b_n) = b_n > 0$. According to the intermediate value theorem, there exists $t_0 \in (0, b_n)$ so that $L(t_0) = 0$, or $H^n(t_0) = 1 - t_0$. Let $t_k = H^k(t_0) \in (0, b_{n-k})$. Then

$$0 < t_0 < t_1 < \cdots < t_{n-1} < t_n < 1.$$

For $k = 1, 2, \ldots, n-1$, from $H(t_k) = t_{k+1}$ we have $G(t_k) = F(t_{k+1})$. When $k = n$,

$$G(t_n) = G(H^n(t_0)) = G(1 - t_0) = G(t_0) = F(H(t_0)) = F(t_1).$$

Now define $g(x) = (f(1 + c) - f(c))G(x - c) + f(c)$, and $a_k = t_k + c$ for $k = 1, 2, \ldots, n$. Then $g(x)$ and $a_1, a_2, \ldots, a_n$ meet the given conditions.

Therefore, $m = 2$ is the required smallest number. $\square$

Comment   The number of contestants who can solve this problem correctly is far smaller than expected.

3   Those points in the plane rectangular coordinate system with both the x- and y- coordinates being rational numbers are called rational points. For any rational point $P(x, y)$, if the product xy is a multiple of 2, but not a multiple of 3, then dye P in red; if xy is a multiple of 3 but not a multiple of 2, then P is colored blue; otherwise, P is colored yellow.

Question: Is there a line segment on the coordinate plane with exactly 58 red rational points and 2017 blue rational points (ignoring yellow points)? Prove your conclusion. (Contributed by Wang Bin)

Solution   As a line segment is part of a straight line, let's first consider the coloring of rational points on a straight line. Assume there are at least two rational points on line l. It is easy to see that l satisfies a linear equation $ax + by = c$, where a, b, c are integers, and a, b are not all 0. Then on l, a red point requires the coordinate product to be an integer of type $6k \pm 2$, and a blue one requires the coordinate product to be an integer of type $6k + 3$.

Since the products of the point coordinates on a coordinate axis are always zero, line l is not a coordinate axis. If l is parallel to a coordinate axis, we may assume $a = 0, bc \neq 0$, so the line is $y = \dfrac{c}{b}$. The red points on the line are $\left((6k \pm 2)\dfrac{b}{c}, \dfrac{c}{b} \right)$, and the blue points are $\left((6k + 3)\dfrac{b}{c}, \dfrac{c}{b} \right)$, showing the arrangement order of "red, blue, red, red, blue, red,..." , which cannot meet the requirements.

Assume $ab \neq 0$. Then a point $P(x, y)$ on l with xy being an integer is equivalent to $ab \mid (ax)(by) = (ax)(c - ax)$. So (ax) and $(c - ax)$ must be integers, since their sum and product are both integers. Therefore, on l we just need consider rational points $(x, y) = \left(\dfrac{z}{a}, \dfrac{c - z}{b} \right)$ with $z = ax$ being an integer.

We first consider set $A_0 = \{z \in \mathbb{Z}, ab \mid z(c - z)\}$, which corresponds to all the rational points on l with their products of the coordinates being integers.

Suppose ab is factored into $ab = 2^{\alpha-1}3^{\alpha_0}p_1^{\alpha_1}\cdots p_r^{\alpha_r}$. For the time being, we consider the case without the factors of 2 and 3, i.e. consider set $A = \{z \in \mathbb{Z} : p_1^{\alpha_1}\cdots p_r^{\alpha_r} \mid z(c-z)\}$.

For $i = 1, 2, \ldots, r$, assume $\upsilon_{p_i}(c) = \gamma_i$ (if $c = 0$, then let $\upsilon_p(0) = +\infty$). We have:

If $\gamma_i \geq \dfrac{\alpha_i}{2}$, then $p_i^{\alpha_i} \mid z(c-z) \Longleftrightarrow p_i^{\left\lceil \frac{\alpha_i}{2} \right\rceil} \mid z$ and at this time we define $D_i = p_i^{\left\lceil \frac{\alpha_i}{2} \right\rceil}$.

If $\gamma_i < \dfrac{\alpha_i}{2}$, then $p_i^{\alpha_i} \mid z(c-z) \Longleftrightarrow p_i^{\gamma_i} \mid z$ and $\dfrac{z}{p_i^{\gamma_i}} \equiv 0$ or $\dfrac{c}{p_i^{\gamma_i}} (\mathrm{mod}\, p_i^{\alpha_i - 2\gamma_i})$, and at this time we define $D_i = p_i^{\gamma_i}$, $\beta_i = \alpha_i - 2\gamma_i$.

We may assume that $\gamma_i < \dfrac{\alpha_i}{2}$ for $i = 1, 2, \ldots, m$, and $\gamma_i \geq \dfrac{\alpha_i}{2}$, for $i = m + 1, \ldots, r$. Then $z \in A$ is equivalent to $D_i \mid z$ (for $i = 1, \ldots, m, m + 1, \ldots, r$) and $\dfrac{z}{D_i} \equiv 0$ or $\dfrac{c}{D_i} (\mathrm{mod}\, p_i^{\beta_i})$ (for $i = 1, 2, \ldots, m$). Therefore, $z = D_1 \times \cdots \times D_r \times T$, where $T \equiv 0$ or $f_i (\mathrm{mod}\, p_i^{\beta_i})$, with f_i satisfying $f_i \times D_1 \cdots D_{i-1}D_{i+1}\cdots D_r \equiv \dfrac{c}{D_i} (\mathrm{mod}\, p_i^{\beta_i})$, for $i = 1, 2, \ldots, m$.

Let's determine all Ts meeting the condition above. Define $M = p_1^{\beta_1}p_2^{\beta_2}\cdots p_m^{\beta_m}$. For each $k \in \{1, 2, \ldots, m\}$, consider the following congruence equations:

$$T_k \equiv f_k(\mathrm{mod}\, p_k^{\beta_k}) \text{ and } T_k \equiv 0(\mathrm{mod}\, p_i^{\beta_i}) \text{ for } i \in \{1, \ldots, m\} - \{k\}.$$

Its solution is

$$T_k \equiv g_k p_1^{\beta_1}\cdots p_{k-1}^{\beta_{k-1}}p_{k+1}^{\beta_{k+1}}\cdots p_m^{\beta_m} = \frac{g_k}{p_k^{\beta_k}}M(\mathrm{mod}\, M),$$

where g_k satisfies

$$g_k p_1^{\beta_1}\cdots p_{k-1}^{\beta_{k-1}}p_{k+1}^{\beta_{k+1}}\cdots p_m^{\beta_m} \equiv f_k(\mathrm{mod}\, p_k^{\beta_k}) \text{ and } 0 < g_k < p_k^{\beta_k}.$$

Letting $I = \{i \mid T \equiv f_i(\mathrm{mod}\, p_i^{\beta_i}), 1 \leq i \leq m\}$, then for $j \in I' = \{1, \ldots, m\} - I$, we have $T \equiv 0 \ (\mathrm{mod}\, p_j^{\beta_j})$. So $T \equiv \sum_{i \in I} T_i(\mathrm{mod}\, p_k^{\beta_k}), \forall k \in \{1, \ldots, m\}$, i.e. $T \equiv \sum_{i \in I} T_i(\mathrm{mod}\, M)$ or

$$T = \varepsilon_1 \frac{g_1}{p_1^{\beta_1}}M + \varepsilon_2 \frac{g_2}{p_2^{\beta_2}}M + \cdots + \varepsilon_m \frac{g_m}{p_m^{\beta_m}}M + \varepsilon_0 M,$$

where $\varepsilon_1, \ldots, \varepsilon_m \in \{0, 1\}, \varepsilon_o \in \mathbb{Z}$. Define A_T as the set of all Ts in these forms. Then $A = \{z \in \mathbb{Z} : p_1^{\alpha_1}\cdots p_r^{\alpha_r} \mid z(c-z)\}$ is the result of multiplying separately all the numbers in A_T by the number in $D = D_1 \times \cdots \times D_r$.

Set A represents those rational points on line l, the denominators of whose coordinate products contain only 2 and 3 prime factors.

Next we need to identify the red dots and the blue dots, that is, the rational points whose coordinate products are integers of $6k \pm 2$ and $6k + 3$ types, respectilvely; and then judge whether it is possible that, on a certain line segment, the number of blue dots is much larger than that of red dots.

The problem has two possible directions: If there is such a line segment, we need to try to make the red and blue dots in a certain interval orderly; if it does not exist, we need to find some core constraints and use the disparity ratio of red and blue dots to introduce contradiction.

Suppose that when do this, we realize that a positive answer to the original question may exist. At this time, we must try to construct an example to confirm it. The purpose of the construction is to *order* the red and blue dots, so the example should be as concise as possible. For this purpose, we set $\alpha_{-1} = \alpha_0 = 0$. Then $ab = p_1^{\alpha_1} \cdots p_r^{\alpha_r}$ is coprime with 2, 3, and set $A = A_0$ represents all rational points on l whose products of coordinates are integers, For the convenience of construction, we further assume that $6 \,|\, c$, i.e., $2, 3 \,|\, z(c - z) \Leftrightarrow 2, 3 \,|\, z$, so that the red point is equivalent to $z = 6k \pm 2$, and the blue one is equivalent to $z = 6k + 3$. (If $c = 6k \pm 2$, it is fine too, although a little more troublesome; but if c is an odd number, there will be no blue dots on the line).

In this way, under the settings of $(ab, 6) = 1$ and $6 \,|\, c$, the number of red and blue dots on a certain line segment is equal to the $6k \pm 2$ type and $6k + 3$ type numbers contained in a certain interval of set A, respectively. Since A is all the numbers in A_T multiplied by $D = D_1 \times \cdots \times D_r$, and D is coprime with 6, so the numbers of red and blue points are equal to that of $6k \pm 2$ type and $6k + 3$ type, respectively, contained in A_T in a certain interval.

$$A_T = \left\{ \left(\varepsilon_0 + \varepsilon_1 \frac{g_1}{p_1^{\beta_1}} + \cdots + \varepsilon_m \frac{g_m}{p_m^{\beta_m}} \right) M : \varepsilon_1, \ldots, \varepsilon_m \in \{0, 1\}, \varepsilon_0 \in \mathbb{Z} \right\},$$

$$\text{①}$$

where $M = p_1^{\beta_1} p_2^{\beta_2} \cdots p_m^{\beta_m}$ is coprime with 6.

We consider the remainders modulo 6 of the elements in A_T: if we assume each of $\dfrac{g_1}{p_1^{\beta_1}}, \ldots, \dfrac{g_m}{p_m^{\beta_m}}$ is smaller than $\dfrac{1}{m}$, then the elements of A_T in interval $[kM, kM + M - 1]$ are those with $\varepsilon_0 = k$ in ①, i.e.,

$$kM + \varepsilon_1 \frac{g_1}{p_1^{\beta_1}} M + \cdots + \varepsilon_m \frac{g_m}{p_m^{\beta_m}} M, \varepsilon_1, \ldots, \varepsilon_m \in \{0, 1\}^m.$$

So there are totally 2^m elements. If we further assume that $g_1 = g_2 = \cdots = g_m = 6$, then all these 2^m elements are of the same modulo 6, i.e., all congruence with kM. At this time , the elements with the same modulo 6 remainder in A_T appear in the same interval, as shown in the following:

$[0, M)$: 2^m elements with $T \equiv 0$ (mod 6);
$[M, 2M)$: 2^m elements with $T \equiv 1$ (mod 6);
$[2M, 3M)$: 2^m elements with $T \equiv 2$ (mod 6), red points;
$[3M, 4M)$: 2^m elements with $T \equiv 3$ (mod 6), blue points;
$[4M, 5M)$: 2^m elements with $T \equiv 4$ (mod 6), red points; and
$[5M, 6M)$: 2^m elements with $T \equiv 5$ (mod 6); $\cdots$

Therefore, we just need take $m \geq 11$, find m prime numbers $p_1, p_2, \ldots, p_m$, all greater than $6m$; and then let

$$a = 1, b = p_1 \times \cdots \times p_m, c = 6 \left(\frac{1}{p_1} + \frac{1}{p_2} + \cdots + \frac{1}{p_m} \right) \times p_1 \times \cdots \times p_m.$$

At this time, $r = m, D_1 = \cdots = D_m = 1, \beta_1 = \cdots = \beta_m = 1, f_1 = \cdots = f_m = c, g_1 = \cdots = g_m = 6$.

In this way, on line $l : ax + by = c$, there are exactly 2^m blue points and no red ones for $x \in [3b, 4b - 1]$, and there are exactly 2^m red points and no blue ones for $x \in [4b, 5b - 1]$. Consequently, there exists a segment on l between $x \in [3b, 5b - 1]$, on which there are exactly 58 red points and 2017 blue ones. $\qquad \square$

Remark If we know the above structure in advance and then write it out directly, the previous discussion could be much simpler.

For any arrangement of red and blue dots, there is a certain line segment on the coordinate plane, corresponding to this arrangement. Because $g_1, \ldots, g_m$, can be designed according to the red and blue needs ($g_i = 2$ for red dots, $g_i = 3$ for blue dots), to make the first m elements of set A_T in interval $(0, M)$

$$\frac{g_1}{p_1^{\beta_1}} M < \frac{g_2}{p_2^{\beta_2}} M < \cdots < \frac{g_m}{p_m^{\beta_m}} M$$

are exactly in the required red-blue order. In fact, at this time, it is necessary to additionally satisfy the magnitude relationship between some prime powers $p_i^{\beta_i}$, but this is easily obtained by the prime number theorem $\pi(x) \sim \frac{x}{\ln x}$, that is, for a sufficiently large interval $[L, (1 + \varepsilon)L]$, there exists always a prime number.

④ Given $0 < l \leq 2$, on the unit circumference with two fixed points C and D, let the length of the moving string AB be l, and the vertices

A, B, C, D of the non-degenerate quadrilateral $ABCD$ are arranged clockwise on the circumference. Line segment AC intersects BD at point P. Find the trajectories of the cirmcumcenters of $\triangle ABP$ and $\triangle BCP$, respectively. (Contributed by Yao Yijun)

Solution   Let the unit circle center be O, the circumcenters of $\triangle ABP$ and $\triangle BCP$ be O_1 and O_2, respectively, the length of CD be $m, \alpha = \arcsin\dfrac{l}{2}$, and $\beta = \arcsin\dfrac{m}{2}$. Note that in the unit circle, the circumferential angle (and the corresponding inferior arc) of a fixed-length chord is fixed, so by the calculation formula of the inner angle of a circle, it can be seen that when string AB continuously changes on the same side of line $CD, \angle CPD = \angle APB$ is fixed, so $\angle AO_1B = 2\angle APB$ is fixed, and $\triangle AO_1B$ is an isosceles triangle with fixed angle A. At this time, the trajectory of O_1 is a circular arc with O as the center.

Similarly, when string AB continuously changes on the same side of $CD, \angle BO_2C = 2\angle BPC$ is fixed, and $\triangle BO_2C$ is an isosceles triangle with a fixed vertex angle. At this time, the trajectory of O_2 is the image of an arc on the unit circle after the rotation homothetic transformation centered on C.

Finally, the condition "$A, B, C,$ and D are arranged clockwise on the circumference" determines that we must discuss the following situations.

Case 1: The path from C to D clockwisely on the unit circle is a minor arc.

Case 1.1: If $l < m$, then the possible positions of B are an arc on the unit circle (the corresponding center angle is $2\pi - 2\beta - 2\alpha$, with C being one of the endpoints). At this time, the trajectory of O_1 is an arc centered on O, with the radius of $\cos\alpha - \sin\alpha\cot(\alpha + \beta)$ and the center angle of $2\pi - 2\beta - 2\alpha$ that is symmetrical with the perpendicular bisector of CD. The trajectory of O_2 is obtained in this way: taking the arc that is the trajectory of B, first perform a homothetic transformation centered on C, using $\dfrac{1}{2\sin(\beta + \alpha)}$ as the coefficient, and then rotate $\dfrac{\pi}{2}(\beta + \alpha)$ clockwisely.

Case 1.2: If $l > m$, note that there are two possible positions for AB, depending on whether the center O and line CD are on the same side or opposite sides of AB.

(i) If the center O and line CD are on the same side of AB, then we get the same result as discussed above.

(ii) If the center O and line CD are on the opposite sides of AB, then the possible positions of B are an arc on the unit circle (the corresponding center angle is $2\alpha - 2\beta$, with C being one of the endpoints). At this time, the trajectory of O_1 is an arc centered on O, with the radius of $\sin\alpha\cot(\alpha - \beta) - \cos\alpha$ and the center angle of $2\alpha - 2\beta$ that is symmetrical with the perpendicular bisector of CD. The trajectory of O_2 is obtained in this way: taking the arc that is the trajectory of B, first perform a homothetic transformation centered on C, using $\dfrac{1}{2\sin(\alpha - \beta)}$ as the coefficient, and then rotate $\dfrac{\pi}{2}(\alpha - \beta)$ anti-clockwisely.

Case 2: The path from C to D clockwisely on the unit circle is a major arc.

Case 2.1 If $l < m$, then the possible positions of B are an arc on the unit circle (the corresponding center angle is $2\beta - 2\alpha$, with C being one of the endpoints). At this time, the trajectory of O_1 is an arc centered on O, with the radius of $\cos\alpha + \sin\alpha\cot(\beta - \alpha)$ and the center angle of $2\beta - 2\alpha$ that is symmetrical with the perpendicular bisector of CD. The trajectory of O_2 is obtained in this way: taking the arc that is the trajectory of B, first perform a homothetic transformation centered on C, using $\dfrac{1}{2\sin(\beta - \alpha)}$ as the coefficient, and then rotate $\dfrac{\pi}{2} - (\beta - \alpha)$ anti-clockwisely.

Case 2.2 If $l > m$, then there are no positions for A and B on the unit circle. So the set of trajectories of O_1, O_2 is empty. $\qquad\square$

Remark On the last day of the Selection Test, we needed a relatively simple geometry problem, so we picked one from an old book of geometry problem set (published in France in 1920), and made a slight change of it. Because everyone in the coaching staff couldn't remember any problem involving the concept of trajectory had been tested in previous domestic competitions, we thought this problem was relatively safe from being familiar by the contestants.

5 For any real numbers x, y, z, a real coefficient polynomial with 3 variables satisfies

$$f(x, y, z) = f_2(x, y)z^2 + 2f_3(x, y)z + f_4(x, y) \geq 0,$$

where $f_k(x, y)$ is a kth homogeneous polynomial ($k = 2, 3, 4$). Suppose there exists a real coefficient polynomial $r(x, y)$, such that

$$f_2(x, y)f_4(x, y) - f_3^2(x, y) = (r(x, y))^2.$$

Prove: there are two real coefficient polynomials $g(x, y, z)$ and $h(x, y, z)$, satisfying

$$f(x, y, z) = g^2(x, y, z) + h^2(x, y, z).$$

(Contributed by Han Jingjun)

Solution If $0 \neq f_2 = l^2$ is the square of a polynomial of the first degree, then from $l^2 f_4 \geq f_3^2$ we have $l \mid f_3, l \mid r$. Let $f_3 = lt$, $r = ls$. Then $f_4 - t^2 = s^2$, so $f = (lz + t)^2 + s^2$, meeting the requirements.

If f_2 is strictly positive, then f_2 is irreducible and can be written as the sum of two squares of first-order polynomials, i.e., $f_2 = l_1^2 + l_2^2$. Note that

$$l_1^2 + l_2^2 \mid (f_3^2 + r^2)l_1^2 - f_3^2(l_1^2 + l_2^2) = (l_1 r + l_2 f_3)(l_1 r - l_2 f_3)$$

Then either $f_2 \mid l_1 r + l_2 f_3$ or $f_2 \mid l_1 r - l_2 f_3$. Consequently, for the following two groups of equations about h_1 and h_2,

$$\begin{cases} h_1 = \dfrac{f_3 l_1 - r l_2}{f_2}, \\ h_2 = \dfrac{r l_1 + f_3 l_2}{f_2}, \end{cases} \text{ and } \begin{cases} h_1 = \dfrac{f_3 l_1 + r l_2}{f_2}, \\ h_2 = \dfrac{-r l_1 + f_3 l_2}{f_2}, \end{cases} \qquad \text{(1)}$$

there is one group in which h_2 is a polynomial, and from

$$h_1^2 + h_2^2 = \frac{(l_1^2 + l_2^2)(f_3^2 + r^2)}{f_2^2} = \frac{f_3^2 + r^2}{f_2} = f_4,$$

we know h_1 is also a polynomial. Furthermore,

$$l_1 h_1 + l_2 h_2 = f_3.$$

Therefore, there exist polynomials h_1, h_2, such that

$$F = (l_1^2 + l_2^2)z^2 + 2(l_1 h_1 + l_2 h_2)z + h_1^2 + h_2^2 = (l_1 z + h_1)^2 + (l_2 z + h_2)^2.$$

The proof is completed. $\qquad \square$

Remark It can be seen from the proof that the method to solve this question is constructive. In 1888, D. Hilbert proved that a ternary quartic homogeneous nonnegative polynomial can be written as the sum of three squares of real coefficient polynomials (Hilbert, D. Über die darstellung definiter formen als summe von formenquadraten. Mathematische Annalen, 1888,32(3): 342–350. But there has not been a constructive proof for this comclusion. In fact, Hilbert proved that if and only if $n \leq 2$ and d is even, or $d = 2$, or $(n, d) = (3, 4)$, a real coefficient nonnegative homogeneous polynomial of degree d with n variables can be expressed as the sum of squares of polynomials.

This question is closely related to the Hilbert 1888 Theorem. Using the ideas of this question, we can give an elementary constructive proof of a special case of the Hilbert 1888 Theorem: if a ternary quartic homogeneous nonnegative polynomial f has real zeros, then f can be written as the sum of three squares of real coefficient polynomials.

Hilbert gave his famous lecture on "Mathematical Problems" at the 2nd World Congress of Mathematicians held in Paris in 1900. He raised 23 mathematical problems, of which the 17th one was about the sum of squares. That is, can a semi-positive definite polynomial with real coefficients be expressed as the sum of several squares of several rational functions with real coefficients? In 1927, Artin solved the 17th problem of Hilbert on the basis of establishing a real field theory. He proved that a real coefficient semi-positive definite polynomial can always be expressed as the sum of several squares of real coefficient rational functions. However, Artin's proof is not constructive, and so far people have not obtained a full constructive proof of the Hilbert 17 problem.

6 We say a simple undirected graph G has a flowing k-coloring, if it satisfies:

(1) we can put a piece with one of k colors on each vertex of G, and the pieces on any two adjacent vertexes are with different colors;

(2) in each round we can take a Hamilton circle in G (the same Hamilton circle can be reused) $(v_1, v_2, \ldots, v_n)$, moving the piece on $v_i (i = 1, 2, \ldots, n-1)$ to v_{i+1}, and that on v_n to v_1, with each pair of adjacent pieces still in different color;

(3) after a finite of rounds, each piece has reached every vertex.

Let $T(G)$ be the smallest k, so that G has a flowing k-coloring; if there is no such k, let $T(G) = -1$; $\chi(G)$ represents the color number of G.

Question: for what integer $\eta \geq 2$, there is a graph G, so that G does not contain cycles shorter than 2017, $\chi(G) \leq \eta$, and $T(G) \geq 2^\eta$? (Contributed by Chen Xiaomin)

Solution The answer is $\eta \geq 3$.

Firstly, we show that any graph G with $\chi(G) \leq 2$ has $T(G) \leq 2 < 2^2$. If $\chi(G) = 1$, G does not have any edge, then either $T(G) = -1$ or $T(G) = 1$ (when G has only one point). If $\chi(G) = 2$, G is a bipartite graph. At this time, if G does not include a Hamiltonian circle, then $T(G) < 0$. Otherwise, take any Hamiltonian circle H in G, and dye the vertexes on it with black and white colors, such that all edges are between black and white vertexes, and the colors of all vertexs on H is black-white spaced. So the number of vertexes on G is an even number (let it be $2k$), and the vertexs on H are in order

$$(a_1, a_2, \ldots, a_{2k}).$$

The subscripts of the two vertexes of each edge in G are of opposite parity. If i is an odd number, we put a red piece on a_i, otherwise a blue piece. Then let all pieces move along H in each round. It is obvious that, after each round of movement, the colors of the adjacent two pieces are still different, and after $2k$ rounds, all the pieces have visited all the vertices. So $T(G) \leq 2$.

Next we prove that for any given positive integer C, there exists a graph G such that G does not contain a cycle whose length is less than 2017, $\chi(G) \leq 3$ and $T(G) > C$. We construct G as follows: first take a circle of N (to be determined) vertexes C_N with the vertexes arranged in order $(a_1, a_2, \ldots, a_N, a_1)$. Then let $D = 2017C + 1$, and construct a sequence $x_0 = 1, x_i = x_{i-1} + 2017 + i, i = 1, \ldots, D$. Let $N > x_D$, and connect $a_{x_i} a_{x_{i+1}}, i = 1, \ldots, D - 1$. We prove that G satisfies the following condition:

(1) G does not include circles shorter than 2017. This is obvious from our construction and the fact that $N > x_D, D > 2017$.

(2) $\chi(G) \leq 3$. Consider the order of vertexes $a_1, \ldots, a_N$, each vertex is only adjacent to at most two ones in front of it (particularly, a_N is only adjacent to a_{N-1} and a_1, as $N > x_D$). So we can dye every vertex in this order with three colors, only need to make sure that the color each vertex used is different from that used by the neighbors in front of it.

(3) $T(G) \neq -1$. Since C_N is a Hamiltonian circle in G, G has at least one flow N-coloring: put a piece with a color different from the others on each vertex, and each round the pieces move along the C_N.

(4) C_N is the only Hamiltonian circle in G. It is easy to see that if a vertex has a degree of 2, then any Hamiltonian circle will pass through all edges connected to it. For any edge $a_i a_{i+1}$ in C_N, from our construction, at least one of a_i and a_{i+1} has a degree of 2, so any Hamiltonian circle will contain edge $a_i a_{i+1}$. Similarly, the degree of a_N is 2, so any Hamiltonian will contains edge $a_N a_1$.

(5) $T(G) > C$. From (4), all pieces in each round can only move clockwise or counterclockwise along C_N, so the order of them on C is unchanged after each round. Assume at the begining the piece on a_i is P_i, for any $i \leq x_{D-1}$. Then for $d < D, P_i$ and $P_{i+2017+d}$ are in different colors. Otherwise, by the condition of the question, P_i will reach x_d in certain rounds, and at this time $P_{i+2017+d}$ reaches $x_d + 2017 + d = x_{d+1}$, but by our construction, the two vertexes are adjacent in G. In general, we have that, for any $1 \leq i < j \leq D, j - i \geq 2017, P_i$ and P_j are in different colors. Therefore, the number of occurrences for any color in $P_1, P_2, \ldots, P_D$ will not exceed 2016, so the total number of colors is at least $D/2016 > C$. $\qquad\square$

Remark For any graph G, we can easily add edges to construct a graph G' such that $\chi(G) = \chi(G')$, and G' has a unique Hamiltonian circle. According to Erdös' famous theorem: there exists a graph with the minimum cycle length greater than 2017 and $\chi > \eta$. So if we change the condition $\chi(G) \leq \eta$ in the question to $\chi(G) = \eta$, it is also true.

China National Team Selection Test

2018

The first round of the 2018 IMO China National Team Selection Test was held from December 28, 2017 to January 10, 2018 at the No.1 Middle School Affiliated to Central China Normal University in Hubei Province. The main task was to select Chinese national team members to participate in the 59th International Mathematical Olympiad in Romania in 2018. A total of 59 players participated in the first round. After two tests (with equal weights), 15 players were selected to enter the second round.

The second round was held in the School of Mathematical Sciences of East China Normal University and the Zizhu Campus of the No.2 Middle School Affiliated to East China Normal University in Shanghai, from March 18 to March 28, 2018. During this period, two tests were conducted, and the total scores of the four tests were compared (with equal weights). The top six scorers were identified to be the Chinese national team members for the 59th IMO. The six players were Li Yixiao, Yao Rui, Chen Yiyi, Ye Qi, Wang Zeyu and Ouyang Zexuan.

The coaches of the national training team are:

Qu Zhenhua (East China Normal University);

Xiong Bin (East China Normal University);

Yu Hongbing (Suzhou University);

Leng Gangsong (Shanghai University);

Li Ting (Sichuan University);

Ai Yinghua (Tsinghua University);

Wang Bin (Institute of Mathematics and Systems Science, Chinese Academy of Sciences);

He Yijie (East China Normal University);

Fu Yunhao (Guangdong Second Normal University);

Zhang Sihui (University of Shanghai for Science and Technology).

Test I, First Day
(8:00 – 12:30; December 30, 2017)

1 Given real numbers $p, q > 0$ and $p + q = 1$, show that for every 2017-tuple $(y_1, y_2, \ldots, y_{2017})$ of real numbers, there exists a unique 2017-tuple $(x_1, x_2, \ldots, x_{2017})$ of reals, satisfying

$$p \max\{x_i, x_{i+1}\} + q \min\{x_i, x_{i+1}\} = y_i, i = 1, 2, \ldots, 2017,$$

where $x_{2018} = x_1$. (Contributed by Liu Zhipeng)

Solution　First, we consider just one equation and show for arbitrary real numbers x and y, there exists a unique real number $z = f_y(x)$, satisfying

$$p \max\{x, z\} + q \min\{x, z\} = y. \qquad \text{①}$$

There are two cases:

　　Case 1: $y \geq x$. If $z < x$, then

$$p \max\{x, z\} + q \min\{x, z\} = px + qz < px + qx = x \leq y,$$

and ① cannot hold. We have $z \geq x$,

$$p \max\{x, z\} + q \min\{x, z\} = pz + qx = y,$$

and $z = \dfrac{y - qx}{p}$. Indeed, $z = \dfrac{y - qx}{p} \geq \dfrac{x - qx}{p} = x$ is true.

　　Case 2: $y < x$. If $z > x$, then

$$p \max\{x, z\} + q \min\{x, z\} = pz + qx > px + qx = x > y,$$

and ① cannot hold. We have $z \leq x$,

$$p \max\{x, z\} + q \min\{x, z\} = px + qz = y,$$

and $z = \dfrac{y - px}{q}$. Indeed, $z = \dfrac{y - px}{q} \leq \dfrac{x - px}{q} = x$ is true.

Therefore, $f_y(x)$ is given by

$$f_y(x) = \begin{cases} \dfrac{y - qx}{p}, & x \le y, \\ \dfrac{y - px}{q}, & x > y. \end{cases}$$

Now consider the system of 2017 equations in the original problem. Essentially, we are required to show there exists a unique 2017-tuple $(x_1, x_2, \ldots, x_{2017})$, such that $x_{i+1} = f_{y_i}(x_i)$, $i = 1, 2, \ldots, 2017$.

Since x_{i+1} is determined solely by x_i, it is equivalent to showing the equation

$$x_1 = (f_{y_{2017}} \circ f_{y_{2016}} \circ \cdots \circ f_{y_1})(x_1)$$

has a unique solution x_1. Notice that for fixed y, $f_y(x)$ is a continuous, strictly decreasing function in x; the composition of 2017 such functions is also continuous and strictly decreasing (since 2017 is odd). Define

$$F(x) = (f_{y_{2017}} \circ f_{y_{2016}} \circ \cdots \circ f_{y_1})(x) - x.$$

Then $F(x)$ is continuous and strictly decreasing, with

$$\lim_{x \to \infty} F(x) = +\infty, \quad \lim_{x \to +\infty} F(x) = -\infty.$$

By the intermediate value theorem, there exists a unique $x_1 \in R$ satisfying $F(x) = 0$, completing the proof. $\square$

2 A positive integer is called an "interesting number", if the number of its positive factors is divisible by 2018. Find all positive integers d, such that there exists an infinite arithmetic sequence of interesting numbers, and the common difference of the sequence is equal to d. (Contributed by Wang Bin)

Solution $2018 = 2 \times 1009$, let $q = 1009$ be prime. For a positive integer n, denote $\tau(n)$ as the number of positive factors of n. Evidently, n is an interesting number if and only if $2 \mid \tau(n)$ and $q \mid \tau(n)$, which holds if and only if n is not a perfect square, and there exists a prime number p, such that

$$v_p(n) \equiv -1 \pmod{q},$$

where $v_p(n)$ is defined as the power of p in n.

Lemma *Let a, d be positive integers. If $q \mid \tau(a + nd)$ for every positive integer n, then there exists a prime p, such that $v_p(d) \ge q$.*

Proof of lemma   By contradiction, assume for every prime p, $v_p(d) < q$. Let $(a, d) = k, a = ka', d = kd'$, and

$$d' = p_1^{\alpha_1} \cdots p_m^{\alpha_m}, k = p_1^{\beta_1} \cdots p_m^{\beta_m} p_{m+1}^{\gamma_1} \cdots p_{m+l}^{\gamma_l},$$

in which $1 \le \alpha_i \le q - 1$, $0 \le \beta_i \le q - 1 - \alpha_i \le q - 2$, $1 \le i \le m$; $1 \le \gamma_j \le q - 1$, $1 \le j \le l$. Since a' and d' are coprime, and $p_1 \cdots p_m \mid d'$, we deduce that $p_1 \cdots p_m$ and a' are coprime, $a' + nd'$ and $p_1 p_2 \cdots p_m$ are also coprime. Let $b = p_1^{\beta_1} \cdots p_m^{\beta_m}$, $c = p_{m+1}^{\gamma_1} \cdots p_{m+l}^{\gamma_l}$, $k = bc$. We have

$$\tau(a + nd) = \tau(b)\tau[c(a' + nd')]$$

$$= (\beta_1 + 1) \cdots (\beta_m + 1)\tau[c(a' + nd')].$$

From $0 < \beta_i + 1 < q$, $1 \le i \le m$, it follows $q \mid \tau(a + nd) \Leftrightarrow q \mid \tau[c(a' + nd')]$.

Consider terms in $\{c(a' + nd')\}$ that are divisible by $(p_{m+1} \cdots p_{m+l})^q$, which is equivalent to requiring

$$a' + nd' \equiv 0 \pmod{p_{m+1}^{q-\gamma_1} \cdots p_{m+l}^{q-\gamma_l}}.$$

As d' and $p_{m+1} \cdots p_{m+l}$ are coprime, the above linear congruence equation has solutions which are given by the arithmetic sequence

$$n = n_0 + p_{m+1}^{q-\gamma_1} \cdots p_{m+l}^{q-\gamma_l} t.$$

Consider its subsequence

$$c[a' + (n_0 + p_{m+1}^{q-\gamma_1} \cdots p_{m+l}^{q-\gamma_l} t)d'] = (p_{m+1} \cdots p_{m+l})^q (a_1 + td'),$$

in which

$$a_1 = \frac{a' + n_0 d'}{p_{m+1}^{q-\gamma_1} \cdots p_{m+l}^{q-\gamma_l}}.$$

Note that $a' + n_0 d'$ and d' are coprime, hence a_1 and d' are coprime as well.

Since $q \mid \tau[(p_{m+1} \cdots p_{m+l})^q (a_1 + td')]$ holds if and only if for some prime p,

$$v_p[(p_{m+1} \cdots p_{m+l})^q (a_1 + td')] \equiv -1 \pmod{q},$$

but

$$v_p[(p_{m+1} \cdots p_{m+l})^q (a_1 + td')] \equiv v_p(a_1 + td') \pmod{q},$$

then we deduce $q \mid v_p(a_1 + td') + 1$. For each t, let $p = p(t)$ be a prime such that $q \mid v_p(a_1 + td') + 1$. Clearly, $p^{q-1} \mid a_1 + td'$, and so $p \le (a_1 + td')^{\frac{1}{q-1}}$. As $a_1 + td'$ and d' are coprime, p and d' are coprime as well.

Let N be a positive integer. Consider $p(1), p(2), \ldots, p(N)$: if $p(i) = p(j)$, then $p^{q-1} \mid j - i$. Consequently for a prime p, p may appear at most $\dfrac{N}{p^{q-1}} + 1$ times in $p(1), p(2), \ldots, p(N)$. On the other hand, by the given condition, these p's appear totally N times. Then we have

$$N \leq \sum_{p \leq (a_1 + Nd')^{\frac{1}{q-1}}} \left(\frac{N}{p^{q-1}} + 1 \right) \leq \frac{3}{4}N + (a_1 + Nd')^{\frac{1}{q-1}},$$

where the second inequality is derived from

$$\sum_p \frac{1}{p^{q-1}} \leq \sum_{i=2}^{\infty} \frac{1}{i^2} \leq \frac{1}{4} + \sum_{i=3}^{\infty} \frac{1}{i(i-1)} = \frac{3}{4}.$$

When N is sufficiently large, $N < N$, leading to a contradiction. The lemma is then proved.

Return to the original problem. Let E be the set of all d's, each d being the common difference of an infinite arithmetic sequence of interesting numbers. By the lemma there exists a prime p, such that $v_p(d) \geq q$.

If $d = 2^q$, the arithmetic sequence $a + 2^q n$ does not meet the requirement: indeed, if $v_2(a) \geq q$, then $a + 2^q n = 2^q(2^{-q}a + n)$; since q is odd, we can take n such that $2^{-q}a + n = 2m^2$; then $a + 2^q n$ is a perfect square, which is clearly a contradiction. If $0 < v_2(a) = \alpha < q$, then $a + 2^q n = 2^\alpha(a' + 2^{q-\alpha}n), 2 + a'$. By the lemma, there exists n, such that $q + \tau(a' + 2^{q-\alpha}n)$; but $q \mid \tau(a + 2^q n)$, hence $\alpha = q - 1$. Taking m such that $a' + 2m$ is a perfect square, we have $a + 2^q m = 2^{q-1}(a' + 2m)$ also as a perfect square, a contradiction again. If $v_2(a) = 0$, a is odd, then by the proof of the lemma ($a_1 = a$, $d' = 2^q$), we see that $q \mid \tau(a + 2^q n)$ cannot hold for every n. Hence $d = 2^p \notin E$.

For the other situations, the desired sequence always exists. For an odd prime p, take r as a quadratic nonresidue modulo p, and r is odd. There are three cases:

(1) $a_n = (pn + r)p^{q-1}$, satisfying $v_p(a_n) = q - 1$, and $pn + r$ is a quadratic nonresidue modulo p. Clearly, $pn + r$ is not a perfect square, and nor is a_n. The common difference is $d = p^q$. By taking a subsequence, we may obtain all situations with $p^q \mid d$.

(2) $a_n = (2pn + r)2^{q-1}$, satisfying $v_2(a_n) = q - 1$, and $2pn + r$ is a quadratic nonresidue modulo p. Likewise, neither $2pn + r$ nor a_n is a perfect square, and $d = 2^q \cdot p$. Taking a subsequence, and by the arbitrariness of p, we may obtain all situations with $2^q \mid d, d$ not a power of 2.

(3) $a_n = (4n + 3)2^{q-1}$, satisfying $v_2(a_n) = q - 1$, and $4n + 3, a_n$ are not perfect squares, and $d = 2^{q+1}$. Taking a subsequence, we may obtain all situations with $2^{q+1} \mid d$.

In summary, $d \in E$ if and only if there exists a prime p, such that $v_p(d) \geq 1009$, and $d \neq 2^{1009}$. □

Remark The proof may be substantially simplified by applying the Dirichlet theorem: when a and d are coprime, $\{a + nd\}_{n \geq 1}$ includes infinitely many prime numbers. In the proof of the lemma, there exists t such that $a_1 + td'$ is a prime, thereby $q + \tau(a_1 + td') = 2$. In the discussion $d = 2^q$, if $v_2(a) = \alpha < q$, then there exists n such that $a' + 2^{q-\alpha}n$ is a prime, and we must have $\alpha = q - 1$.

③ In $\triangle ABC$, circle ω touches sides AB, AC at D, E, respectively, with $D \neq B$, $E \neq C$, and $BD + CE < BC$. Let points F, G be on side BC, with $BF = BD$, $CG = CE$. Let DG, EF meet at K. Point L lies on the minor arc $\overset{\frown}{DE}$ of ω, such that the tangent of ω at L is parallel to BC. Prove that the incenter of $\triangle ABC$ lies on line KL. (Contributed by Lin Tianqi)

Solution 1 As shown in Fig. 3.1, let I be the incenter of $\triangle ABC$. We need to prove that L, I, K are collinear.

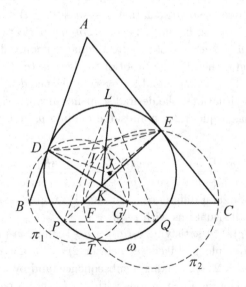

Fig. 3.1

Let π_1 and π_2 be the circumcircles of $\triangle BDG$ and $\triangle CEF$, respectively. Clearly,

$$IF = ID = IE = IG,$$

and hence D, E, F, G are concyclic (let ω_1 be their circumcircle, centered at J), and we have

$$\angle IDB + \angle IGB$$
$$= (180° - \angle IDA) + \angle IFC$$
$$= 180°.$$

This implies that I is on π_1. Likewise, I is on π_2. Let T be the intersection of π_1 and π_2 other than I. Connect TD, TE, JD, JE, and notice that

$$\angle DTE = \angle DTI + \angle ETI = \angle DBI + \angle ECI$$
$$= \frac{1}{2}(\angle ABC + \angle ACB) = \frac{1}{2}(180° - \angle BAC)$$
$$= \frac{1}{2}\angle DJE.$$

We infer that T is on ω. By applying the radical center theorem to ω_1, π_1 and π_2, we deduce that the radical axes DG, EF, IT are concurrent, and hence I, K, T are collinear. Now we show that L is on this line as well.

Let ray TI and circle ω meet at L_0 (not labeled in the figure as it will be shown $L_0 = L$). Extend EF, DG to meet ω at P, Q, respectively. Connect PQ, PL, QL, IF and IG. From

$$\angle KFG = \angle KDE = \angle KPQ,$$

we have $PQ//FG$. From

$$\angle PL_0T = \angle PET = \angle FET = \angle FIT,$$

we have $PL_0//FI$. Similarly, $QL_0//GI$. This implies that $\triangle L_0PQ$ can be derived from $\triangle IFG$ by a homogeneous dilation. From $IF = IG$, we get $L_0P = L_0Q$, and the tangent of ω at L_0 is parallel to PQ, which is parallel to BC as well since $PQ//BC$. So $L_0 = L$.

Finally, for the homogeneous dilation between $\triangle LPQ$ and $\triangle IFG$, the intersection K of PF and QG is the homothetic center. Therefore, L, I, K are collinear and the proof is complete.

Solution 2 As shown in Fig. 3.2, let l be the tangent of ω at L, and $l//BC$. Let I be the incenter of $\triangle ABC$. Let lines FI, GI intersect l at X, Y, respectively. Connect AI, BI, CI, DI and EI.

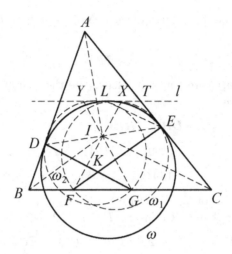

Fig. 3.2

Clearly, $AD = AE$. Since $BD = BF$, $CE = CG$, by symmetry we get $ID = IE = IF = IG$, and hence D, E, F, G are concyclic.

Let line l and AC meet at point T. From $TE = TL$, we have

$$\angle XLE = \angle TLE = \frac{1}{2}\angle ATL = \frac{1}{2}\angle ACB.$$

Moreover, $\angle IEA = \angle IGF = \angle IFG$, and thus C, E, I, F are concyclic. It follows that

$$\angle XFE = \angle IFE = \angle ICE = \frac{1}{2}\angle ABC.$$

Now $\angle XLE = \angle XFE$, and we conclude that E, F, L, X are concyclic, say on circle ω_1. Analogously, D, G, L, Y are concyclic, say on circle ω_2.

Finally, we prove that points K, I and L lie on the radical axis of ω_1 and ω_2; hence they are collinear, and that completes the proof. Obviously, as an intersection point, L does. Denote $p_{\omega_i}(\cdot)$ as the power of a point of circle $\omega_i (i = 1, 2)$. Notice that $l \parallel BC$, and $IF = IG$. It follows $IX = IY$, and

$$p_{\omega_1}(I) = -IF \cdot IX = -IG \cdot IY = p_{\omega_2}(I).$$

Since D, E, F, G are concyclic, it follows

$$p_{\omega_1}(K) = -KE \cdot KF = -KD \cdot KG = p_{\omega_2}(K).$$

We deduce that K, I lie on the radical axis, too.

Solution 3　As shown in Fig. 3.3, let LT be a diameter of ω. Since the tangent of ω at L is parallel to BC, so is the tangent at T: let it meet lines

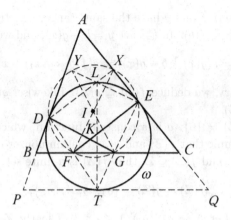

Fig. 3.3

AB, AC at points P, Q, respectively. From $BD = BF$, we have $PD = PT$, $\angle DBF = \angle DPT$, and

$$\triangle BDF \backsim \triangle PDT.$$

Hence D, F, T are collinear. Likewise, E, G, T are collinear.

From $AD = AE, BD = BF$, $CE = CG$, by symmetry we obtain $ID = IE = IF = IG$, and thereby denote $\odot I$ as the circle that passes through D, E, F and G.

Let lines DL, EL meet $\odot I$ at points X, Y (different from D, E), respectively. Since LT is a diameter of ω,

$$\angle FDX = \angle TDL = 90°.$$

It follows that FX is a diameter of $\odot I$, and GY likewise. So FX and GY intersect at I: apply Pascal's theorem to D, E, F, G, X and Y in $\odot I$, and we conclude that K, I, L are collinear. The proof is complete. $\qquad\square$

Test I, Second Day
(8:00 – 12:30; December 31, 2017)

4 Let f and g be two mappings from $\mathbb{Z}$ to $\mathbb{Z}$, satisfying

$$f(g(x) + y) = g(f(y) + x) \qquad (*)$$

for all integers x and y. If f is bounded, prove that g is periodic, i.e., there exists $T > 0$ such that $g(x + T) = g(x)$ for all $x \in \mathbb{Z}$. (Contributed by Qu Zhenhua)

Solution   We assert f and g have the same range, i.e., $f(\mathbb{Z}) = g(\mathbb{Z})$.

Let $a \in f(\mathbb{Z}), a = f(b)$. In (*), let $y = b - g(x)$, and we have

$$a = f(b) = f(g(x) + b - g(x)) = g(f(b - g(x)) + x) \in g(\mathbb{Z}).$$

Since a is arbitrary, we deduce $f(\mathbb{Z}) \subseteq g(\mathbb{Z})$. Likewise, $g(\mathbb{Z}) \subseteq f(\mathbb{Z})$, and hence $f(\mathbb{Z}) = g(\mathbb{Z})$.

If $f(\mathbb{Z}) = g(\mathbb{Z}) = \{0\}$, g is a constant function, which is periodic. In the following, assume that $f(\mathbb{Z})$ and $g(\mathbb{Z})$ contain nonzero integers. Notice that f is bounded and $f(\mathbb{Z}) \subseteq \mathbb{Z}$, thus $f(\mathbb{Z})$ is a finite set. Let

$$f(\mathbb{Z}) = g(\mathbb{Z}) = \{a_1, a_2, \ldots, a_k\},$$

in which $a_1 \neq 0$. Let $A_i = g^{-1}(a_i)$, $1 \leq i \leq k$. Clearly, $\mathbb{Z} = A_1 \cup \cdots \cup A_k$. Suppose that $f(y_0) = a_1$. For any $x_1, x_2 \in A_i$,

$$\begin{aligned}
g(a_1 + x_1) = g(f(y_0) + x_1) &= f(g(x_1) + y_0) \\
&= f(g(x_2) + y_0) = g(f(y_0) + x_2) \\
&= g(a_1 + x_2).
\end{aligned}$$

It follows that all elements in $a_1 + A_i$ belong to A_j, for some $1 \leq j \leq k$. This is true for every $1 \leq i \leq k$, and that is to say, $j = j(i)$ is uniquely determined by $a_1 + A_i \subseteq A_j$. Consider the mapping $j : \{1, 2, \ldots, k\} \to \{1, 2, \ldots, k\}$, which satisfies $a_1 + A_i \subseteq A_{j(i)}$ for every $1 \leq i \leq k$. For every $1 \leq t \leq k$, take $x \in A_t - a_1$, $x \in A_i$. By $a_1 + x \in A_t$, $j(i) = t$, we deduce that j is surjective, and hence bijective.

Suppose that j divides $\{1, 2, \ldots, k\}$ into m cycles, with lengths $l_1, l_2, \ldots, l_m$. If A_i lies in a cycle of length l, then for $x \in A_i$, $x + la_1 \in A_i$. We have

$$g(x + la_1) = g(x) = a_i, \forall x \in A_i.$$

For $T = l_1 l_2 \cdots l_m a_1$, T is a multiple of every la_1, and it follows $g(x) = g(x + T)$ for every $x \in \mathbb{Z}$. Therefore, g is a periodic function with period T.   □

5   Given a positive integer k. For a positive integer n, if the number of binomial coefficients among $C_n^0, C_n^1, \ldots, C_n^n$ that are divisible by k is at least $0.99n$, then n is called "good". Prove there exists a positive integer N, such that the number of good integers among $1, 2, \ldots, N$ is at least $0.99N$. (Contributed by Fu Yunhao)

Solution First, we need a lemma.

Lemma *Let p be a prime, m be a positive integer, and $\varepsilon > 0$. There exists a positive integer A, such that for any positive integer $a \geq A$, the number of binomial coefficients among $\mathrm{C}_x^y (x, y \in \{0, 1, 2, \ldots, a\})$ that are not divisible by p^m is at most $\varepsilon(a+1)^2$.*

Proof of lemma Let positive integers M and u be sufficiently large, so that inequalities $\left(\dfrac{3}{4}\right)^M mM^m < \dfrac{\varepsilon}{2}$ and $\left(\dfrac{u}{u+1}\right)^2 > 1 - \dfrac{\varepsilon}{2}$ hold. We claim that $A = up^M$ meets the requirement.

For any $a \geq A$, let $a + 1 = tp^M + r$, in which $0 \leq r < p^M$, $t \geq u$. The number of binomial coefficients among $\mathrm{C}_x^y (0 \leq x, y \leq a)$ with $x, y \in \{0, 1, \ldots, tp^M - 1\}$ is

$$(tp^M)^2 = (a+1)^2 \left(\frac{tp^M}{a+1}\right)^2 \geq (a+1)^2 \left(\frac{tp^M}{(t+1)p^M}\right)^2$$

$$= (a+1)^2 \left(\frac{t}{t+1}\right)^2 \geq (a+1)^2 \left(\frac{u}{u+1}\right)^2$$

$$> (a+1)^2 \left(1 - \frac{\varepsilon}{2}\right).$$

Now represent x and y in base-p, adding zeros to the front if necessary, so that both representations have at least M digit places. Then $x = (\cdots x_{M-1} \cdots x_0)_p$, $y = (\cdots y_{M-1} \cdots y_0)_p$, $x_i, y_i \in \{0, 1, \ldots, p - 1\}$. Define S as the set of (x, y)'s with the property: among the last M digit places, there are at least m ones (i's) satisfying $x_i < y_i$. Observe that for each digit place i, there are $\dfrac{1}{2}p(p + 1)$ combinations with $x_i \geq y_i$, and $\dfrac{1}{2}p(p - 1)$ combinations with $x_i < y_i$. Thus the number of (x, y) pairs, which each have, in their last M digit places, exactly l ones (i's) satisfying $x_i < y_i$, is given by

$$t^2 (p^M)^2 \mathrm{C}_M^l \left(\frac{\frac{1}{2}p(p-1)}{p^2}\right)^l \left(\frac{\frac{1}{2}p(p+1)}{p^2}\right)^{M-l}.$$

We have

$$|S| = t^2 p^{2M} \left(1 - \sum_{l=0}^{m-1} C_M^l \left(\frac{\frac{1}{2}p(p-1)}{p^2} \right)^l \left(\frac{\frac{1}{2}p(p+1)}{p^2} \right)^{M-l} \right)$$

$$\geq t^2 p^{2M} \left(1 - \left(\frac{3}{4} \right)^M \sum_{l=0}^{m-1} C_M^l \right)$$

$$\geq t^2 p^{2M} \left(1 - \left(\frac{3}{4} \right)^M m M^m \right)$$

$$\geq t^2 p^{2M} \left(1 - \frac{\varepsilon}{2} \right)$$

$$\geq (a+1)^2 \left(1 - \frac{\varepsilon}{2} \right)^2$$

$$\geq (a+1)^2 (1 - \varepsilon).$$

By Kummer's theorem, p^m divides C_x^y if $(x, y) \in S$. It follows that the number of binomial coefficients C_x^y that are not divisible by p^m is at most $\varepsilon(a+1)^2$, and the proof is complete.

Return to the original problem. Let $k = p_1^{\alpha_1} \cdots p_s^{\alpha_s}$ be the prime factorization of k. By the lemma, there exists a positive integer N, such that for $n \geq N$, the number of $C_x^y (0 \leq x, y \leq n)$ that are not divisible by some $p_i^{\alpha_i}$ is at most $\dfrac{(n+1)^2}{8 \times 10^6}$, so the number of them not divisible by k is at most $\dfrac{(n+1)^2}{8 \times 10^6}$. If the number of good numbers among $1, 2, \ldots, n$ is less than $0.99n$, at least $0.01n$ of them are not good: suppose they are $x_1, x_2, \ldots, x_q, q \geq 0.01n$. Then the number of $C_x^y (0 \leq x, y \leq n)$ that are not divisible by k is at least

$$0.01(x_1 + x_2 + \cdots + x_q) \geq 0.01 \frac{q(q+1)}{2} \geq \frac{n^2}{2 \times 10^6} \geq \frac{(n+1)^2}{8 \times 10^6},$$

a contradiction. This implies for $n \geq N$, among $1, 2, \ldots, n$, no fewer than $0.99n$ of them are good numbers. $\square$

6 Let m, n be positive integers, $A_1, A_2, \ldots, A_m$ be m subsets of an n-set. Prove

$$\sum_{i=1}^{m} \sum_{j=1}^{m} |A_i| \cdot |A_i \cap A_j| \geq \frac{1}{mn} \left(\sum_{i=1}^{m} |A_i| \right)^3, \tag{*}$$

where $|X|$ denotes the number of elements of X. (Contributed by Ai Yinghua)

Solution Let $A_1, A_2, \ldots, A_m \subseteq X$, $|X| = n$. If $A_i = \emptyset$, by removing A_i the LHS (left hand side) of (*) is unchanged but the RHS (right hand side) becomes larger as m decreases If $A_1 \cup A_2 \cup \cdots \cup A_m \neq X$, replacing X by $A_1 \cup A_2 \cup \cdots \cup A_m$, the LHS of (*) is unchanged, but the RHS becomes larger as n decreases. Hence we may assume that all A_i's are nonempty, and $A_1 \cup A_2 \cup \cdots \cup A_m = X$.

Let $Y = \{A_1, A_2, \ldots, A_m\}$, and $G = < X, Y, E >$ be the bipartite graph such that for $x \in X$, $A_i \in Y$, $xA_i \in E$ if $x \in A_i$. Consider the set

$$S = \{(x, A_i, y, A_j) | x, y \in X, A_i, A_j \in Y, xA_i, yA_i, yA_j \in E\}.$$

We calculate $|S|$. For fixed A_i, A_j, there are $|A_i|$ ways of choosing x, and $|A_i \cap A_j|$ ways of choosing y. Hence

$$|S| = \sum_{i=1}^{m} \sum_{j=1}^{m} |A_i| \cdot |A_i \cap A_j|.$$

On the other hand, for fixed A_i, y, there are $\deg(A_i)$ ways of choosing x, and $\deg(y)$ ways of choosing A_j, where $\deg(A_i)$, $\deg(y)$ represent the respective degrees of A_i, y in G. Hence

$$|S| = \sum_{yA_i \in E} \deg(A_i)\deg(y).$$

Since all A_i's are nonempty, and $A_1 \cup A_2 \cup \cdots \cup A_m = X$, we have $\deg(A_i) \neq 0$, $\deg(y) \neq 0$. Recall that

$$m = \sum_{yA_i \in E} \frac{1}{\deg(A_i)}, n = \sum_{yA_i \in E} \frac{1}{\deg(y)}.$$

By the generalized Hölder's inequality, we have

$$mn \left(\sum_{i=1}^{m} \sum_{j=1}^{m} |A_i| \cdot |A_i \cap A_j| \right)$$

$$= \left(\sum_{yA_i \in E} \frac{1}{\deg(A_i)} \right) \left(\sum_{yA_i \in E} \frac{1}{\deg(y)} \right) \left(\sum_{yA_i \in E} \deg(A_i)\deg(y) \right)$$

$$\geq \left(\sum_{y A_i \in E} 1 \right)^3 = |E|^3$$

$$= \left(\sum_{i=1}^{m} \deg(A_i) \right)^3$$

$$= \left(\sum_{i=1}^{m} |A_i| \right)^3.$$

The inequality (*) is yielded. □

Test II, First Day
(8:00 – 12:30; January 8, 2018)

1. Given $\triangle ABC$ on the plane α, let D, E, F be moving points on sides BC, CA, AB, respectively, such that $BD = CE$, $CD = BF$. Let P be the intersection point of the circle through B, D, F and the circle through C, D, E other than D. Prove that there exists a point Q on α, such that the length of PQ is constant. (Contributed by He Yijie)

Solution As shown in the figure, take X, Y on the perpendicular bisector of segment BC, such that BX, CY bisect $\angle ABC$, $\angle ACB$, respectively. Clearly, $XB = XC$, $\angle XBF = \angle XBC = \angle XCB$. By $BF = CD$, it follows that $\triangle XBF$ and $\triangle XCD$ are congruent, $\angle BFX = \angle CDX$, and hence B, F, X, D are concyclic. Likewise, C, E, Y, D are concyclic.

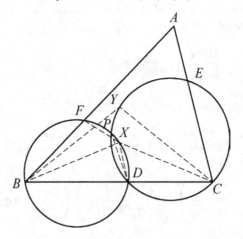

If $AB = AC$, then $X = Y, P$ coincides with X. It follows that P is a fixed point in α, and the length of PQ is constant for any $Q \neq P$.

If $AB \neq AC$, then $X \neq Y$. We define directed angle $\angle(\ell_1, \ell_2)$ as the smallest angle that line ℓ_1 is rotated counterclockwise to be parallel to line ℓ_2. The sum and difference of directed angles are taken modulo π.

As B, D, P, X are concyclic, $\angle(PX, PD) = \angle(BX, BD)$. As C, D, P, Y are concyclic,

$$\angle(PD, PY) = \angle(CD, CY).$$

It follows that

$$\begin{aligned}
\angle(PX, PY) &= \angle(PX, PD) + \angle(PD, PY) \\
&= \angle(BX, BD) + \angle(CD, CY) \\
&= \angle(BX, CY).
\end{aligned}$$

As $\angle(BX, CY)$ is constant, P lies on a fixed circle through X and Y. Taking Q as the center of this circle, the length of PQ is constant. $\qquad\square$

2 Let $P(n)$ represent the number of possible (unordered) partitions of positive integer n. For instance, $P(4) = 5$, since there are 5 partitions as follows:

$$4, 1 + 3, 2 + 2, 1 + 1 + 2, 1 + 1 + 1 + 1.$$

Find all positive integers n, satisfying

$$P(n) + P(n + 4) = P(n + 2) + P(n + 3). \tag{*}$$

(Contributed by Qu Zhenhua)

Solution For convenience, denote P_n as the set of all possible partitions of n, in which every partition π is written as $(a_1 + a_2 + \cdots + a_k)$ in the ascending order, i.e. $a_1 \leq a_2 \leq \cdots \leq a_k$, denote $m(\pi)$ as the smallest part a_1, and $l(\pi)$ as the length k.

Let $Q(n + 1) = P(n + 1) - P(n)$ be the first order difference of the partition function. Then (*) becomes

$$Q(n + 4) = Q(n + 2) + Q(n + 1). \tag{**}$$

To better understand the combinatorial meaning of $Q(n + 1)$, consider the injective mapping

$$\varphi : P_n \to P_{n+1}, (a_1 + \cdots + a_k) \mapsto (1 + a_1 + \cdots + a_k).$$

The image of φ is $\varphi(P_n) = \{\pi \in P_{n+1} \,|\, m(\pi) = 1\}$, and hence

$$Q(n + 1) = |P_{n+1} \backslash \varphi(P_n)| = |\{\pi \in P_{n+1} \,|\, m(\pi) \geq 2\}|.$$

Denote $Q_{n+1} = \{\pi \in P_{n+1} \,|\, m(\pi) \geq 2\}$, $Q(n + 1) = |Q_{n+1}|$. To understand $Q(n + 4) - Q(n + 2)$ in (**), consider the injective mapping

$$\psi : Q_{n+2} \to Q_{n+4}, (a_1 + \cdots + a_k) \mapsto (2 + a_1 + \cdots + a_k).$$

The image of ψ is $\psi(Q_{n+2}) = \{\pi \in Q_{n+4} \,|\, m(\pi) = 2\}$, and hence

$$Q(n + 4) - Q(n + 2) = |Q_{n+4} \backslash Q_{n+2}| = |\{\pi \in Q_{n+4} \,|\, m(\pi) \geq 3\}|.$$

Denote $R_{n+4} = \{\pi \in Q_{n+4} \,|\, m(\pi) \geq 3\}$. It suffices to find all n, such that $|R_{n+4}| = |Q_{n+1}|$.

Divide partitions in each of R_{n+4} and Q_{n+1} into four categories:

$$R_{n+4} = R_{n+4}^1 \cup R_{n+4}^2 \cup R_{n+4}^3 \cup R_{n+4}^{\geq 4},$$
$$Q_{n+1} = Q_{n+1}^1 \cup Q_{n+1}^2 \cup Q_{n+1}^3 \cup Q_{n+1}^{\geq 4},$$

where the superscripts 1, 2, 3 and ≥ 4 stand for the length of the partitions in that category. Clearly,

$$|R_{n+4}^1| = |Q_{n+1}^1| = 1.$$

Consider the mapping

$$f : Q_{n+1}^2 \to R_{n+4}^2, (a_1 + a_2) \mapsto (b_1 + b_2),$$

in which $b_1 = a_1 + 1 \geq 3$, $b_2 = a_2 + 2$. Then f is injective, and its image is $f(Q_{n+1}^2) = \{(b_1 + b_2) \in R_{n+4}^2 \,|\, b_1 < b_2\}$. Consequently, $R_{n+4}^2 \backslash f(Q_{n+1}^2) =$

$\{(a + a)\}$ if and only if n is even, here $a = \dfrac{n}{2} + 2 \geq 3$. Let

$$\delta_n = |R^2_{n+4}| - |Q^2_{n+1}| = \begin{cases} 1, & 2\,|\,n, \\ 0, & 2n. \end{cases}$$

The mapping

$$g : R^3_{n+4} \to Q^3_{n+1}, (a_1 + a_2 + a_3) \mapsto (b_1 + b_2 + b_3)$$

is a bijection, in which $b_i = a_i - 1, i = 1, 2, 3$. Hence $|R^3_{n+4}| = |Q^3_{n+1}|$. Finally, consider the mapping

$$h : R^{\geq 4}_{n+4} \to Q^{\geq 4}_{n+1}, (a_1 + a_2 + a_3 + a_4 + \cdots + a_k)$$
$$\mapsto (b_1 + b_2 + b_3 + a_4 + \cdots + a_k)$$

where $b_i = a_i - 1, i = 1, 2, 3$. Evidently, h is injective, and the image $h(R^{\geq 4}_{n+4})$ consists of all partitions in $Q^{\geq 4}_{n+1}$ of the form $(a_1 + a_2 + a_3 + a_4 + \cdots + a_k)$ with $a_3 < a_4$. Denote

$$H_{n+1} = Q^{\geq 4}_{n+1} \backslash h(R^{\geq 4}_{n+4}) = \{(a_1 + \cdots + a_k) \in Q^{\geq 4}_{n+4}\,|\,a_3 = a_4\}.$$

Then

$$|Q_{n+1}| - |R_{n+4}| = (|Q^1_{n+1}| - |R^1_{n+4}|) + (|Q^2_{n+1}| - |R^2_{n+4}|)$$
$$+ (|Q^3_{n+1}| - |R^3_{n+4}|) + (|Q^{\geq 4}_{n+1}| - |R^{\geq 4}_{n+4}|)$$
$$= |H_{n+1}| - \delta_n.$$

We must find all n, such that $|H_{n+1}| = \delta_n$, which implies (**). Taking some small n values and partitions with $l(\pi) \geq 4, m(\pi) \geq 2, a_3 = a_4$:

$$(2 + 2 + 2 + 2) \in H_8, (2 + 2 + 3 + 3) \in H_{10},$$
$$(2 + 2 + 2 + 2 + 2) \in H_{10}, (2 + 3 + 3 + 3) \in H_{11},$$
$$(2 + 2 + 2 + 2 + 3) \in H_{11}, \ldots.$$

We can see that

(1) if $n \geq 7$, n is odd, H_{n+1} is nonempty, $(2 + 2 + \cdots + 2) \in H_{n+1}$, but $\delta_n = 0$, not good;

(2) if $n = 1, 3, 5, H_{n+1} = \varnothing, \delta_n = 0$, as desired;

(3) if $n = 2, 4, 6, H_{n+1} = \varnothing, \delta_n = 1$, not good;

(4) if $n \geq 10, n$ is even, $|H_{n+1}| \geq 2$. Indeed, for $x = n - 7 \geq 3$, $y = \frac{1}{2}(n - 4) \geq 3$,

$$(2 + 2 + 2 + 2 + x) \in H_{n+1}, (2 + 3 + y + y) \in H_{n+1},$$

but $\delta_n = 1$, not good;

(5) if $n = 8, H_9 = \varnothing, \delta_8 = 1$, not good;

In summary, if (*) holds, n must equal 1, 3, or 5.   □

3 Let p, q be fixed positive integers. Betty plays a solitary game on the blackboard. Initially, there are n positive integers on the board. In every turn, Betty may erase two identical integers a and a, and write $a + p, a + q$ on the board. If all the integers on the board are distinct, the game is over. Find the smallest n for which Betty may play on the n integers infinitely. (Contributed by Wang Bin)

Solution The smallest n equals $\dfrac{p + q}{(p, q)}$.

First, consider the case that $(p, q) = 1$. We may assume $p \leq q$. Define A as a multiset (i.e., an element can appear in the set more than once) of the n integers on the blackboard, and $L(a)$ as the operation that replaces two a's in A by $a + p$ and $a + q$. If $A = \{1, 2, \ldots, p, 1, 2, \ldots, q\}$, after applying $L(1)$, we obtain

$$\{2, 3, \ldots, p + 1, 2, 3, \ldots, q + 1\} = A + 1.$$

Clearly, after getting $A + k$, we may apply $L(1 + k)$ to get $A + (k + 1)$.

For necessity, we must show that if the game can be played infinitely, then $n \geq p + q$. Among all such multisets, let A_0 be one with the fewest integers, $|A_0| = n$. Let $L(a_1), L(a_2), \ldots$ be the infinite sequence of operations on A_0 and accordingly we obtain $A_1, A_2, \ldots$.

Notice that in A_0 the smallest number $\min A_0$ must appear at least twice: otherwise, no operation can be applied to $\min A_0$, and Betty can play with the other $n - 1$ numbers infinitely, contradicting the choice of n.

Let $m = \min A_0$. If m does not appear in the sequence $a_1, a_2, \ldots$, then Betty can play with the other $n - 2$ numbers infinitely, again it is a contradiction. So we may assume $a_k = m, k$ being the minimal. Furthermore, we may rearrange the operation sequence $L(a_1), L(a_2), \ldots, L(a_k)$ to be $L(a_k), L(a_1), \ldots, L(a_{k-1})$, because the two sequences have the same effect. It follows that we may let $a_1 = \min A_0$, and similarly let $a_2 = \min A_1, a_3 =$

min A_2, and so on, so every operation is exerted on the smallest number in the set.

Consider $D_k = \max A_k - \min A_k$. If $D_k > q$, since A_{k+1} is obtained from A_k through $L(\min A_k)$, we have $\max A_{k+1} = \max A_k, \min A_{k+1} \geq \min A_k$, and hence

$$D_{k+1} \leq D_k.$$

If $D_k \leq q$, then $\max A_{k+1} = \min A_k + q, \min A_{k+1} \geq \min A_k$, and hence

$$D_{k+1} \leq q.$$

From the above argument, we can see there exists a positive integer M, such that $D_k \leq M$ for all $k \geq 0$. Therefore, all numbers in the multiset $A_k - (\min A_k)$ must be from $0, 1, 2, \ldots, M$. Evidently, there are only finitely many ways to choose n numbers from $M+1$ candidates, so there exist $s < t$, such that

$$A_s - (\min A_s) = A_t - (\min A_t),$$

which means $A_t = A_s + m$ for some positive integer m. Let every multiset $B = \{b_1, b_2, \ldots, b_n\}$ correspond to a polynomial

$$f_B(x) = x^{b_1} + x^{b_2} + \cdots + x^{b_n}.$$

If A_{k+1} is obtained from A_k through $L(a)$, then

$$f_{A_{k+1}}(x) = f_{A_k}(x) - 2x^a + x^{a+p} + x^{a+q} \equiv f_{A_s}(x) \pmod{x^p + x^q - 2}.$$

Hence $f_{A_t}(x) \equiv f_{A_s}(x) \pmod{x^p + x^q - 2}$. Moreover, by $A_t = A_s + m$ we have $f_{A_t}(x) = x^m f_{A_s}(x)$. It follows

$$x^p + x^q - 2 \mid (x^m - 1) f_{A_s}(x).$$

We claim that the greatest common divisor of $x^p + x^q - 2$ and $x^m - 1$ in $\mathbb{Z}[x]$ is $x - 1$. Let η be a complex root of $x^m - 1$. If η is also a root of $x^p + x^q - 2$, from $|\eta^p| = |\eta^q| = 1$, we have $\eta^p = \eta^q = 1$. Since $(p, q) = 1$, by Bézout's theorem there exist u, v, such that $pu + qv = 1$, and thus $\eta = \eta^{pu+qv} = 1$, which implies that $x^p + x^q - 2$ and $x^m - 1$ have only one common root $x = 1$. Since $x^m - 1$ is primitive with no repeated root, the greatest common divisor is indeed $x - 1$. In $\mathbb{Z}[x]$, we have

$$\frac{x^p + x^q - 2}{x - 1} = 1 + x + \cdots + x^{p-1} + 1 + x + \cdots + x^{q-1} \mid f_{A_s}(x),$$

or written as

$$f_{A_s}(x) = g(x)(1 + x + \cdots + x^{p-1} + 1 + x + \cdots + x^{q-1})$$

for some $g(x) \in \mathbb{Z}[x]$. Letting $x = 1$, we get $(p+q) \mid f_{A_s}(1) = |A_s| = n, n \geq p + q$.

For general p, q, let $d = (p, q)$. Divide integers in A_0 into d subsets $A_{0,r}, r = 1, 2, \ldots, d$, such that all integers in $A_{0,r}$ are congruent to r modulo d. Notice that if Betty operates on two integers in $A_{0,r}$, the resulting integers are still congruent to r modulo d, and hence they are in $A_{0,r}$. So there is a multiset $A_{0,r}$ that Betty can play infinitely. When n is minimal, $A_0 = A_{0,r}$ for some r. Now we examine the multiset $B_0 = \dfrac{1}{d}(A_0 - r)$ with $p' = \dfrac{p}{d}, q' = \dfrac{q}{d}$, and conclude $|A_0| = |B_0| \geq p' + q' = \dfrac{p+q}{(p,q)}$. $\square$

Test II, Second Day
(8:00 – 12:30; January 9, 2018)

4 Let $k \geq 1$ be an integer, $k - 1$ is not square-free, and M is a positive integer. Prove: there exists real number $\alpha > 0$, such that for every positive integer n, $[\alpha k^n]$ and M are coprime.

Remark   Positive integer m is not square-free if for some integer $d > 1, d^2 \mid m$; $[x]$ represents the largest integer not greater than x. (Contributed by Yu Hongbing)

Solution   Let $k - 1 = q^2 b$, q being prime. Let the distinct prime factors of M other than q be $p_1, p_2, \ldots, p_m$, and $P = p_1 p_2 \cdots p_m$. We claim there exist positive integers x, y, such that

$$q^2 x + q + 1 = Py. \qquad \text{①}$$

Notice for every $p_i, 1 \leq i \leq m$, $(p_i, q) = 1$; thus the congruence equation

$$q^2 x + q + 1 \equiv 0 \pmod{p_i}$$

has solution(s). Moreover, $p_1, p_2, \ldots, p_m$ are pairwise coprime. By the Chinese remainder theorem, there exists a positive integer x, satisfying

$$q^2 x + q + 1 \equiv 0 \pmod{p_i}, i = 1, 2, \ldots, m.$$

For this x, we have integer y satisfying ①.

Take $\alpha = \dfrac{Py}{q}$. From ①, we have for every integer $n \geq 1$,

$$yPk^n = yP(q^2 b + 1)^n \equiv Py \equiv 1 \pmod{q},$$

and thus

$$[\alpha k^n] = \left[\frac{yPk^n}{q}\right] = \frac{yPk^n - 1}{q}. \qquad ②$$

From

$$yPk^n = yP(q^2 b + 1)^n \equiv yP = q^2 x + q + 1 \equiv q + 1 \pmod{q^2},$$

we deduce that $\dfrac{yPk^n - 1}{q}$ is not divisible by q.

On the other hand, we can infer from ② that for every $p_i, 1 \leq i \leq m$, $p_i \times [\alpha k^n]$ (because $p_i \mid P$). Hence $yPk^n - 1 \equiv -1 \pmod{p_i}$, and $p_i \times [\alpha k^n]$. The conclusion is yielded. □

5 Given positive integers n and k satisfying $n \geq 4k$, find the minimum value $\lambda = \lambda(n, k)$, such that the following inequality holds for any positive reals $a_1, a_2, \ldots, a_n$:

$$\sum_{i=1}^{n} \frac{a_i}{\sqrt{a_i^2 + a_{i+1}^2 + \cdots + a_{i+k}^2}} \leq \lambda \quad (a_{n+j} = a_j, j = 1, 2, \ldots, k.).$$

Here $a_{n+j} = a_j, j = 1, 2, \ldots, k$. (Contributed by Fu Yunhao)

Solution $\lambda(n, k) = n - k$.

First, take $a_i = q^{n-i}, i = 1, 2, \ldots, n, q > 1$. Then for $1 \leq i \leq n - k$,

$$1 > \frac{a_i}{\sqrt{a_i^2 + a_{i+1}^2 + \cdots + a_{i+k}^2}} \geq \frac{a_i}{\sqrt{a_i^2 + ka_{i+1}^2}} \to 1, \text{ as } q \to +\infty;$$

and for $n - k < i \leq n$,

$$0 < \frac{a_i}{\sqrt{a_i^2 + a_{i+1}^2 + \cdots + a_{i+k}^2}} \leq \frac{a_i}{a_1} \to 0, \text{ as } q \to +\infty.$$

It follows that

$$\sum_{i=1}^{n} \frac{a_i}{\sqrt{a_i^2 + a_{i+1}^2 + \cdots + a_{i+k}^2}} \to n - k, \text{ as } q \to +\infty,$$

and hence $\lambda(n, k) \geq n - k$.

Next, we claim that for all positive reals $a_1, a_2, \ldots, a_n$,

$$\sum_{i=1}^{n} \frac{a_i}{\sqrt{a_i^2 + a_{i+1}^2 + \cdots + a_{i+k}^2}} \le n - k. \qquad \text{(1)}$$

A special situation is when $n = 4, k = 1$:

$$\sum_{i=1}^{4} \frac{a_i}{\sqrt{a_i^2 + a_{i+1}^2}} \le 3. \qquad \text{(2)}$$

Lemma   Let $x, y, z, w \in (0, 1)$, and $x + y + z + w > 3$. Then
$$(1 - x^2)(1 - y^2)(1 - z^2)(1 - w^2) < x^2 y^2 z^2 w^2.$$

Proof of lemma   It suffices to show $(1 - x^2)(1 - y^2) < z^2 w^2$, since the other part follows analogously. Due to

$$(1 - x^2)(1 - y^2) \le \left(\frac{2 - x^2 - y^2}{2} \right)^2,$$

it suffices to show $\dfrac{2 - x^2 - y^2}{2} < zw$, or $x^2 + y^2 + 2zw > 2$. Since $z + w \ge 3 - x - y > 1$, we have

$$zw > 1 \cdot (2 - x - y) = 2 - x - y,$$

and

$$x^2 + y^2 + 2zw > x^2 + y^2 + 2(2 - x - y)$$
$$= (x - 1)^2 + (y - 1)^2 + 2 > 2.$$

The lemma is proved.

Now we verify (2). Suppose there exist a_1, a_2, a_3, a_4, such that (2) does not hold. Let

$$x = \frac{a_1}{\sqrt{a_1^2 + a_2^2}}, y = \frac{a_2}{\sqrt{a_2^2 + a_3^2}}, z = \frac{a_3}{\sqrt{a_3^2 + a_4^2}}, w = \frac{a_4}{\sqrt{a_4^2 + a_1^2}}.$$

Clearly, $x, y, z, w \in (0, 1)$, and by the hypothesis $x + y + z + w > 3$. It follows from the lemma that

$$(1 - x^2)(1 - y^2)(1 - z^2)(1 - w^2) < x^2 y^2 z^2 w^2,$$

or

$$\frac{1 - x^2}{x^2} \cdot \frac{1 - y^2}{y^2} \cdot \frac{1 - z^2}{z^2} \cdot \frac{1 - w^2}{w^2} < 1.$$

However,

$$\frac{1 - x^2}{x^2} \cdot \frac{1 - y^2}{y^2} \cdot \frac{1 - z^2}{z^2} \cdot \frac{1 - w^2}{w^2} = \frac{a_2^2}{a_1^2} \cdot \frac{a_3^2}{a_2^2} \cdot \frac{a_4^2}{a_3^2} \cdot \frac{a_1^2}{a_4^2} = 1,$$

a contradiction. So (2) is true.

Next, we claim when $n = 4k$, ⓵ holds. Indeed,

$$\sum_{i=1}^{n} \frac{a_i}{\sqrt{a_i^2 + a_{i+1}^2 + \cdots + a_{i+k}^2}} \leq \sum_{i=1}^{n} \frac{a_i}{\sqrt{a_i^2 + a_{i+k}^2}},$$

combined with ⓶, we have

$$\sum_{i=1}^{n} \frac{a_i}{\sqrt{a_i^2 + a_{i+k}^2}} = \sum_{m=1}^{K} \left(\frac{a_m}{\sqrt{a_m^2 + a_{m+k}^2}} + \frac{a_{m+k}}{\sqrt{a_{m+k}^2 + a_{m+2k}^2}} \right.$$

$$\left. + \frac{a_{m+2k}}{\sqrt{a_{m+2k}^2 + a_{m+3k}^2}} + \frac{a_{m+3k}}{\sqrt{a_{m+3k}^2 + a_m^2}} \right)$$

$$\leq 3k = n - k.$$

So when $n = 4k$, ⓵ is true.

For $n > 4k$, without loss of generality, assume $a_n = \max_{1 \leq i \leq n} a_i$. For $n - k \leq i \leq n - 1$,

$$\frac{a_i}{\sqrt{a_i^2 + \cdots + a_n^2 + \cdots + a_{i+k}^2}}$$

$$\leq \frac{a_i}{\sqrt{a_i^2 + \cdots + a_{n-1}^2 + a_{n+1}^2 + \cdots + a_{i+k}^2 + a_{i+k+1}^2}},$$

while

$$\frac{a_n}{\sqrt{a_n^2 + a_1^2 + \cdots + a_k^2}} \leq 1.$$

Let $b_i = a_i$, $1 \leq i \leq n - 1$, and the subscript of b_i be modulo $n - 1$. We have

$$\sum_{i=1}^{n} \frac{a_i}{\sqrt{a_i^2 + a_{i+1}^2 + \cdots + a_{i+k}^2}} \leq 1 \sum_{i=1}^{n-1} + \frac{b_i}{\sqrt{b_i^2 + b_{i+1}^2 + \cdots + b_{i+k}^2}},$$

which corresponds to $\lambda = 1 + \lambda(n - 1, k)$. Since this is true for all $n > 4k$, we have $\lambda(n, k) = n - 4k + \lambda(4k, k) = n - k$, yielding the asserted result. $\square$

6 Let $a, b, r \in \mathbb{Z}$, $a \geq 2$, $r \geq 2$. Suppose function $f : \mathbb{Z} \to \mathbb{Z}$ and integer M satisfy the following conditions:

(1) For every $n \in \mathbb{Z}$, $f^{(r)}(n) = an + b$, where $f^{(r)}$ represents the r-th iteration of f, i.e., $f^{(r)} = f(f^{(r-1)})$;

(2) for every $n \geq M$, $f(n) \geq 0$;

(3) for every $m > n \geq M$, $(m - n) \mid (f(m) - f(n))$.

Prove $a = c^r$ for some $c \in \mathbb{Z}$. (Contributed by Li Ting)

Solution   Define $f^{(0)}(n) = n$. Since $f^{(r)}(n) = an + b$ is injective, f must be injective, and so is $f^{(k)}$ for all nonnegative k.

Denote $N_0 = N_{-1} = M$. For $n \geq N_0$, $f^{(0)}(n) = n \geq N_{-1}$. Suppose we have

$$N_0 \leq N_1 \leq \cdots \leq N_t,$$

such that for $i = 0, 1, \ldots, t$ and $n \geq N_i$, $f^{(i)}(n) \geq N_{i-1}$. Then for $n \geq N_t$, since $f^{(t)}(n) \geq N_{t-1} \geq M$, by condition (2) we infer that $f^{(t+1)}(n) = f(f^{(t)}(n)) \geq 0$. Moreover, $f^{(t+1)}$ is injective, and $\{0, 1, \ldots, N_t\}$ is finite. So, there exists $N_{t+1} \geq N_t$, such that for $n \geq N_{t+1}$, $f(n) > N_t$. Inductively, we may construct an integer sequence $N_0 \leq N_1 \leq \cdots \leq N_k \leq \cdots$.

Now for $m > n \geq N_{r-1}$ and $i \in \{0, 1, \ldots, r - 1\}$, by condition (3) we have

$$g_i(m, n) := \frac{f^{(i+1)}(m) - f^{(i+1)}(n)}{f^{(i)}(m) - f^{(i)}(n)}$$

as an integer. Furthermore,

$$\prod_{i=0}^{r-1} g_i(m, n) = \frac{f^{(r)}(m) - f^{(r)}(n)}{f^{(0)}(m) - f^{(0)}(n)} = \frac{(am + b) - (an + b)}{m - n} = a.$$

Thus $g_i(m, n) \mid a$. Let $n_0 = N_{r-1}$, $g(m) = g_0(m, n_0) = \dfrac{f(m) - f(n_0)}{m - n_0}$. Since the number of positive and negative factors of a is finite, there must be a factor c such that $K = \{m \in \mathbb{Z} \mid m > n_0, g(m) = c\}$ is an infinite set. It follows for every $m \in K$, $g(m) = c$, and $f(m) = cm + d$, where $d = f(n_0) - cn_0 \in \mathbb{Z}$. By condition (2) and K being infinite, we deduce $c > 0$.

We claim for all $n \geq n_0$, $f(n) = cn + d$. As

$$\lim_{x \to +\infty} \frac{cx + d - f(n)}{x - n} = c,$$

there exists $N' > n_0$, such that when $x > N'$,

$$\left| \frac{cx + d - f(n)}{x - n} - c \right| < \frac{1}{2}. \tag{1}$$

Since K is infinite, there exists $m' \in K$, $m' > N'$. From condition (3) and inequality $\boxed{1}$, we obtain

$$\frac{f(m') - f(n)}{m' - n} = \frac{cm' + d - f(n)}{m' - n} = c,$$

and this implies $f(n) = cn + d$.

Finally, we claim that for every $k \geq 1$, there exist integers n_k, d_k, such that $f^{(k)}(n) = c^k n + d_k$ provided that $n \geq n_k$. Apply induction on k: when $k = 1$, taking $n_1 = n_0$, $d_1 = d$ suffices. Assume the conclusion holds for k. Then there exists $n_{k+1} > n_k$, such that for $n \geq n_{k+1}$, $f^{(k)}(n) = c^k n + d_k \geq n_1$, and

$$f^{(k+1)}(n) = f(f^{(k)}(n)) = f(c^k n + d_k)$$
$$= c(c^k n + d_k) + d = c^{k+1} n + cd_k + d.$$

Taking $d_{k+1} = cd_k + d$, the induction is complete. In particular, for $n \geq n_r$, $f^{(r)}(n) = c^r n + d_r$; by condition (1), $f^{(r)}(n) = an + b$. Hence $a = c^r$. $\square$

Test III, First Day
(8:00 – 12:30; March 20, 2018)

1 Given separating and non-intersecting circles ω_1 and ω_2, centered at O_1 and O_2, with radii r_1 and r_2, respectively, where $r_1 < r_2$, let AB and XY be common internal tangents of ω_1 and ω_2, such that A, X lie on ω_1, and B, Y lie on ω_2. The circle with diameter AB intersects ω_1 and ω_2 at $P, Q (P \neq A, Q \neq B)$ respectively. If $\angle AO_1P + \angle BO_2Q = 180°$, find $\dfrac{PX}{QY}$ as a function of r_1 and r_2.

(Contributed by Xiong Bin and Lin Tianqi)

Solution As shown in the figure, let S be the intersection point of AB and XY, which is also the homothetic center of ω_1 and ω_2, and thus $\dfrac{AS}{BS} = \dfrac{r_1}{r_2}$. Let M be the midpoint of segment AB. Since $\angle APB = \angle AQB = 90°$, $MA = MP = MB = MQ$, and MP, MQ are tangent to ω_1, ω_2, respectively.

It follows that

$$\angle PO_1M = \frac{1}{2}\angle AO_1P = \frac{1}{2}(180° - \angle BO_2Q)$$
$$= 90° - \angle BO_2M = \angle BMO_2.$$

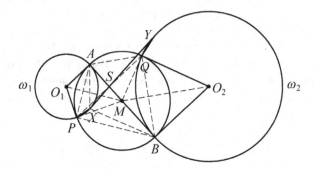

Hence $\triangle O_1PM \backsim \triangle MBO_2$, and this implies

$$\frac{PO_1}{PM} = \frac{BM}{BO_2} = \sqrt{\frac{PO_1}{PM} \cdot \frac{BM}{BO_2}} = \sqrt{\frac{r_1}{r_2}}.$$

On the other hand, from $\angle PAB = \angle PO_1M = \angle BMO_2 = \angle BAQ$, we infer P, Q symmetric about AB. Let AB and PQ intersect at S'. We have

$$\frac{AS'}{S'B} = \frac{S_{\triangle AS'P}}{S_{\triangle PS'B}} = \left(\frac{AP}{BP}\right)^2 = \cot^2\angle PAB = \cot^2\angle PO_1M = \frac{r_1}{r_2} = \frac{AS}{BS}.$$

This indicates $S' = S$, and hence P, S, Q are collinear. By the circle power theorem (or power of a point theorem),

$$SP \cdot SQ = SA \cdot SB = SX \cdot SY.$$

We deduce $\triangle SXP \backsim \triangle SQY$, and thus

$$\frac{PX}{QY} = \frac{SP}{SY} = \frac{SP}{SB} = \frac{AP}{PB} = \sqrt{\frac{r_1}{r_2}}. \qquad \square$$

2 Let G be a simple graph of 100 vertices. Given that for every vertex u, there exists a vertex v adjacent to u, and no vertex is adjacent to both u and v. Find the maximum possible number of edges in G. (Contributed by Fu Yunhao)

Solution   Let $G = (V, E)$. For edge $uv \in E$, if no vertex is adjacent to both u and v, then we call uv "good". Denote E_0 as the set of good edges, and $G_0 = (V, E_0)$. By the hypothesis, every vertex of G is incident to a good edge, i.e., there is no isolated vertex in G_0.

Operation on edges of G_0 as follows: for $uv \in E_0$, if both $\deg(u)$ and $\deg(v)$ in G_0 are not less than 2, then delete uv. Observe that neither u nor v becomes isolated after the operation. Continue this process until we obtain $G_1 = (V, E_1)$, which has no isolated vertex, and for each $uv \in E_1$, $\deg(u) = 1$ or $\deg(v) = 1$ in G_1.

We claim that G_1 is the union of disjoint star graphs. Consider a connected component H of G_1, and H has n vertices. If $n = 2, H = K_{1,1}$. For $n \geq 3$, suppose that uv is an edge of H, with $\deg(u) \geq 2$. Let u be adjacent to $v_1, v_2, \ldots, v_k$. Then $v_1, \ldots, v_k$ all have degree equal to 1 in G_1, and this implies $H = K_{1,k}, k = n - 1$. Moreover, $v_1, v_2, \ldots, v_{n-1}$ are pairwise nonadjacent in G.

Suppose G_1 has m connected components, with $n_1, n_2, \ldots, n_m$ vertices, respectively. Clearly, $n_i \geq 2$, and $\sum_{i=1}^{m} n_i = 100$. We consider edges in G between two stars K_{1,n_i-1} and K_{1,n_j-1}. Let K_{1,n_i-1} have center u and leaves $u_1, \ldots, u_{n_i-1}$, and K_{1,n_j-1} have center v and leaves $v_1, \ldots, v_{n_j-1}$. In G,

(i) If $uv \in E$. Then none of $u_1, \ldots, u_{n_i-1}$ is adjacent to v, and none of $v_1, \ldots, v_{n_j-1}$ is adjacent to u. Hence the number of edges between the two stars cannot exceed $(n_i - 1)(n_j - 1) + 1$.

(ii) If $uv \notin E$, assume u is adjacent to $v_1, \ldots, v_s$ and v is adjacent to $u_1, \ldots, u_t$. Then none of $v_1, \ldots, v_s$ is adjacent to any $u_k, 1 \leq k \leq n_i - 1$, and none of $u_1, \ldots, u_t$ is adjacent to any $v_k, 1 \leq k \leq n_j - 1$. Hence the number of edges between the two stars cannot exceed

$$(n_i - t - 1)(n_j - s - 1) + t + s \leq (n_i - 1)(n_j - 1) + 1.$$

From (i) and (ii), we have

$$|E| \leq \sum_{i=1}^{n}(n_i - 1) + \sum_{1 \leq i < j \leq m}((n_i - 1)(n_j - 1) + 1)$$

$$= \sum_{1 \leq i < j \leq m} n_i n_j - (100 - m)(m - 2)$$

$$= \frac{1}{2}\left(100^2 - \sum_{i=1}^{m} n_i^2\right) - (100 - m)(m - 2)$$

$$\leq 5000 - \frac{5000}{m} - (100 - m)(m - 2).$$

Define $f(m) = 5000 - \dfrac{5000}{m} - (100 - m)(m - 2), 1 \leq m \leq 50$. If $m \geq 17, (100 - m)(m - 2) > 1200$, and $f(m) < 3800$. Evaluating $f(m), 1 \leq m \leq 16$, we find $f(m) \leq 3822$ for $m \neq 8$, and $f(8) = 3823$. But if $|E| = f(8) = 3823$, it must have $n_1 = n_2 = \cdots = n_8$, which cannot be true since $8 \nmid 100$. So $|E| \leq 3822$.

On the other hand, we may construct G as follows. Take four stars $K_{1,11}$, and another four stars $K_{1,12}$, pairwise disjoint. For any two stars, draw an edge between the centers, and draw edges between every pair of leaves on different stars. Then we have $n_1 = n_2 = n_3 = n_4 = 12$, $n_5 = n_6 = n_7 = n_8 = 13$, and

$$|E| = \sum_{i=1}^{8}(n_i - 1) + \sum_{1 \leq i < j \leq 8}((n_i - 1)(n_j - 1) + 1) = 3822.$$

Notice that every edge originally in a star is good in G; hence every vertex is incident to a good edge, and G satisfies the given condition.

So, the maximum possible number of edges in G is 3822.   □

3 Prove: there exists a constant $C > 0$, such that for any positive integer m, and any m positive integers $a_1, a_2, \ldots, a_m$, the inequality

$$H(a_1) + H(a_2) + \cdots + H(a_m) \leq C \left(\sum_{i=1}^{m} i a_{j^i} \right)^{\frac{1}{2}} \qquad ①$$

holds, where $H(n) = \sum_{k=1}^{n} \dfrac{1}{k}$. (Contributed by Yu Hongbing)

Solution 1 It suffices to consider $a_1 \geq a_2 \geq \cdots \geq a_m$. For positive integer j, let s_j denote the number of $a_1, a_2, \ldots, a_m$ that are not less than j. Evidently, $s_j \geq s_{j+1}$; and when $j > a_1$, $s_j = 0$. We have

$$\sum_{i=1}^{m} i a_i = \sum_{i=1}^{m} \sum_{j \leq a_i} i = \sum_{j=1}^{a_1} \sum_{i: a_i \geq j} i = \sum_{j=1}^{a_1} \sum_{i=1}^{s_j} i = \sum_{j=1}^{a_1} \frac{1}{2} s_j(s_j + 1) > \sum_{j=1}^{a_1} \frac{1}{2} s_j^2,$$

and that is

$$\sum_{j=1}^{a_1} s_j^2 \leq 2 \sum_{i=1}^{m} i a_i.$$

On the other hand, let $\delta(j)$ denote the number of $a_1, a_2, \ldots, a_n$ that are equal to j. We have $s_j = \sum_{t \geq j} \delta(t)$, and

$$\sum_{i=1}^{m} H(a_i) = \sum_{j=1}^{a_1} H(j)\delta(j) = \sum_{j=1}^{a_1} \left(\sum_{t \geq j} \delta(j) - \sum_{t \geq j+1} \delta(j) \right) H(j)$$

$$= \sum_{j=1}^{a_1} (s_j - s_{j+1}) H(j) = s_1 H(1) + \sum_{j=2}^{a_1} s_j (H(j) - H(j-1))$$

$$= \sum_{j=1}^{a_1} s_j \cdot \frac{1}{j} \leq \left(\sum_{j=1}^{a_1} s_j^2 \right)^{\frac{1}{2}} \left(\sum_{j=1}^{a_1} \frac{1}{j^2} \right)^{\frac{1}{2}} < \sqrt{\frac{\pi^2}{6}} \left(\sum_{j=1}^{a_1} s_j^2 \right)^{\frac{1}{2}}$$

$$< \sqrt{\frac{\pi^2}{3}} \left(\sum_{i=1}^{m} i a_i \right)^{\frac{1}{2}}.$$

Taking $C = \sqrt{\dfrac{\pi^2}{3}}$, the desired inequality is yielded.

Solution 2 (Collated from Li Yixiao's solution, and simplified by Chen Haoran) First, we need two claims.

Claim 1 *For every positive integer n, $H(n) \leq 1 + \ln n$.*

Proof of Claim 1 When $n = 1$, $H(n) = 1$. When $n > 1$,

$$H(n) = \sum_{k=1}^{n} \frac{1}{k} = 1 + \sum_{k=2}^{n} \frac{1}{k} \leq 1 + \sum_{k=2}^{n} \int_{k-1}^{k} \frac{1}{x} dx$$

$$= 1 + \int_{1}^{n} \frac{1}{x} dx = 1 + \ln n.$$

Claim 2 *For every positive integer n, $\ln(n!) \geq n \ln n - n + 1$. Hence $n! \geq n^n e^{-n+1}$.*

Proof of Claim 2 When $n = 1$, it is obvious. When $n > 1$,

$$\ln(n!) = \sum_{k=2}^{n} \ln k \geq \sum_{k=2}^{n} \int_{k-1}^{k} \ln x dx$$

$$= \int_{1}^{n} \ln x dx = (x \ln x - x) \big|_{1}^{n}$$

$$= n \ln n - n + 1.$$

Return to the original problem. For the left hand side of (1), by Claim 1,

$$\sum_{i=1}^{m} H(a_i) \le m + \sum_{i=1}^{m} \ln a_i = m + \ln(a_1 a_2 \cdots a_m). \tag{*}$$

For the right hand side of (1), apply the AM-GM inequality and by Claim 2, we have

$$C \left(\sum_{i=1}^{m} i a_i \right)^{\frac{1}{2}} \ge C m^{\frac{1}{2}} \left(m!(a_1 a_2 \cdots a_m) \right)^{\frac{1}{2m}}$$

$$\ge C m^{\frac{1}{2}} \left(m^m e^{-m+1} \right)^{\frac{1}{2m}} (a_1 a_2 \cdots a_m)^{\frac{1}{2m}}$$

$$\ge C \cdot \frac{m}{\sqrt{e}} (a_1 a_2 \cdots a_m)^{\frac{1}{2m}}. \tag{**}$$

Let $a_1 a_2 \cdots a_m = x^{2m}, x \ge 1$. To obtain the desired inequality, by (*) and (**), it suffices to find $C > 0$ satisfying

$$1 + 2 \ln x \le \frac{C}{\sqrt{e}} x, \text{ or } \frac{C}{\sqrt{e}} \ge \frac{1 + 2 \ln x}{x} = f(x).$$

Since $f'(x) = 0$ only at $x = \sqrt{e}$, f attains the maximum value at $f(\sqrt{e}) = \frac{2}{\sqrt{e}}$. So (1) holds for $C = 2$. □

Test III, Second Day
(8:00 – 12:00; March 21, 2018)

4 Let $A_1, A_2, \ldots, A_n$ be subsets of $\{1, 2, 3, \ldots, 2018\}$, $|A_i| = 2$ for all i, such that

$$A_i + A_j, 1 \le i \le j \le n$$

are different from each other. Here $A + B = \{a + b \,|\, a \in A, b \in B\}$. Find the largest possible value of n. (Contributed by Qu Zhenhua)

Solution The largest $n = 4033$.

First, we prove $n \le 4033$. If $A_i = A_j, i \ne j$, then $A_i + A_i = A_j + A_j$, contradicting the given condition. Hence $A_1, A_2, \ldots, A_n$ must be different from each other. For $A = \{a, b\}, a < b$, denote $d(A) = b - a$ as the range of A. Suppose there are m distinct values among $d(A_1), d(A_2), \ldots, d(A_n)$, say they are $d_1, d_2, \ldots, d_m$, and there are n_i subsets with range d_i. Clearly, for $i = 1, \ldots, m$, the subsets with range d_i each have different the smallest elements: let set M_i consist of these elements, and $|M_i| = n_i$. Furthermore, let $S_i = \{x - y \,|\, x, y \in M_i, x > y\}$. We assert that $S_1, S_2, \ldots, S_m$ are disjoint. Otherwise, suppose there exist $i \ne j, S_i \cap S_j \ne \emptyset$. We may find x,

$y \in M_i$, $u, v \in M_j$, $x > y$, $u > v$, such that $x - y = u - v$, i.e., $x + v = y + u$. It follows that the four subsets

$$\{x, x + d_i\}, \{y, y + d_i\}, \{u, u + d_j\}, \{v, v + d_j\}$$

satisfy

$$\{x, x + d_i\} + \{v, v + d_j\} = \{y, y + d_i\} + \{u, u + d_j\},$$

contradicting the given condition.

Since the elements of M_i cannot exceed 2017, S_i's must have all elements less than or equal to 2016. As S_i's are disjoint, we have

$$\sum_{i=1}^{m} |S_i| \leq 2016. \qquad \textcircled{1}$$

On the other hand, assume $M_i = \{x_1 < x_2 < \cdots < x_{n_i}\}$. There are at least $n_i - 1$ distinct elements in S_i:

$$x_2 - x_1 < x_3 - x_1 < \cdots < x_{n_i} - x_1,$$

and thus

$$|S_i| \geq n_i - 1, 1 \leq i \leq m. \qquad \textcircled{2}$$

From $\textcircled{1}$, $\textcircled{2}$ and $d(A_i) \leq 2017$, we have $m \leq 2017$, and

$$2016 \geq \sum_{i=1}^{m} |S_i| \geq \sum_{i=1}^{m} (n_i - 1) = n - m.$$

This implies $n \leq m + 2016 \leq 2017 + 2016 = 4033$.

We prove $n = 4033$ is achievable. For $1 \leq i \leq 2017$, let $A_i = \{1, i + 1\}$; for $2018 \leq j \leq 4033$, let $A_j = \{j - 2016, 2018\}$.

(In the following, a simplification is made by Chen Haoran based on the Chinese version.) We call A_k a 1-set, if $1 \in A_k$; a 2018-set, if $2018 \in A_k$. The first 2017 sets are 1-sets, and the last 2017 sets are 2018-sets. For set $A_i + A_j$, there are three possibilities.

(1) The smallest element is 2. Then both A_i and A_j are 1-sets, and $A_i + A_j = \{2, i+1, j+1, i+j+2\}$. No other pairs of sets can give the result.

(2) The largest element is 4036. Then both A_i and A_j are 2018-sets, and similar to (1) the result is unique.

(3) The smallest element is $j - 2015 > 2$ and the largest element is $i + 2019 < 4036$. Clearly the result must be derived from $\{1, i + 1\}$ and $\{j - 2016, 2018\}$.

This completes the proof. $\qquad \square$

5 In $\triangle ABC$, $\angle BAC > 90°$. Let O be the circumcenter, ω be the circumcircle, and the tangent of ω at A meet the tangents at B and C at points P, Q, respectively. Draw perpendiculars from P, Q to side BC, D, E being the feet, respectively. Let points F, G lie on segment PQ different from A, such that A, F, B, E are concyclic, and A, G, C, D are concyclic. Let M be the midpoint of segment DE. Prove that lines DF, OM and EG are concurrent. (Contributed by He Yijie)

Solution As shown in the figure, since A, F, B, E are concyclic and AF is tangent to ω, we have

$$\angle BEF = \angle BAF = \angle ACB,$$

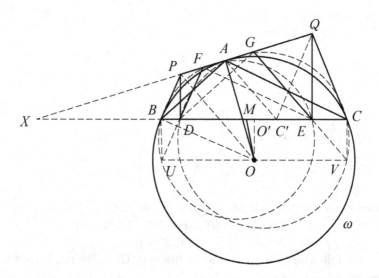

and thus $AC \parallel EF$. Likewise, $AB \parallel DG$. Take point C' on BC, satisfying $C'E = CE$. Then $\angle QC'C = \angle QCC' = \angle PBC$, and thus $PB \parallel QC'$. Without loss of generality, assume $AB < AC$. The rays CB and QP intersect, say at X. From $AB \parallel DG$, it follows

$$\frac{AG}{BD} = \frac{XA}{XB}. \qquad\qquad ①$$

From $\triangle PBD \backsim \triangle QC'E$ and $\triangle XPD \backsim \triangle XQE$, it follows

$$\frac{BD}{C'E} = \frac{BP}{C'Q} = \frac{XB}{XC'}. \qquad\qquad ②$$

Multiply ① by ② to obtain $\dfrac{AG}{C'E} = \dfrac{XA}{XC'}$, and thus $AC' \parallel GE$. Notice Q is the circumcenter of $\triangle ACC'$. We have $\angle BAC' = \angle BAC - \angle C'AC = \angle BAC - \dfrac{1}{2}\angle C'QC = \angle BAC - (90° - \angle QCE) = 90°$. From $AB \parallel DG$ and $AC' \parallel GE$, we infer $\angle DGE = 90°$, and likewise $\angle DFE = 90°$.

Extend FD to meet $\odot ABE$ at U; extend GE to meet $\odot ACD$ at V. As $\angle UBE = \angle UFE = 90°$, $BU \perp BC$. Similarly, $CV \perp BC$. Let perpendicular lines OO' and BC intersect at O'. Evidently, O' is the midpoint of chord BC. Since

$$\angle OBO' = 90° - \angle PBD = \angle BPD,$$

we have $\triangle OBO' \backsim \triangle BPD$, and thus

$$\frac{OO'}{BD} = \frac{OB}{BP} = \cot\angle BOP.$$

By $\angle BOP = \dfrac{1}{2}\angle AOB = \angle ACB$, we get

$$\frac{OO'}{BD} = \cot\angle ACB.$$

In the right $\triangle BDU$, $\dfrac{BU}{BD} = \cot\angle BUD$; as $\angle BUD = \angle BEF = \angle ACB$, we find

$$\frac{BU}{BD} = \cot\angle ACB.$$

Therefore, $OO' = BD$, and $OO' = CV$ likewise. We conclude that $BUVC$ is a rectangle, in which O is the midpoint of side UV. Finally, from $DE \parallel UV$ and M being the midpoint of DE, it follows that lines DF, OM and EG are concurrent. $\qquad\square$

6 Find all positive integer pairs (x, y), satisfying that $(xy+1)(xy+x+2)$ is a perfect square. (Contributed by Mou Xiaosheng)

Solution There is no such pair (x, y). Assume not, and for positive integers $x, y, (xy + 1)(xy + x + 2)$ is a perfect square. Clearly, if $y = 1$,

$(x+1)(2x+2) = 2(x+1)^2$ cannot be a perfect square. So $y > 1$. Let

$$xy + 1 = du^2, xy + x + 2 = dv^2,$$

where d, u, v are positive integers, and d is square free. From

$$dv^2 y - du^2(y+1) = (xy + x + 2)y - (xy + 1)(y+1) = y - 1,$$

it follows that $d \mid y - 1$, and

$$yv^2 - (y+1)u^2 = \frac{y-1}{d}. \qquad \qquad ①$$

However, more generally, for positive integers $y, k, 0 \le k < y$, the Diophantine equation

$$(y+1)u^2 - yv^2 = -k \qquad \qquad ②$$

has no integral solution (u, v). Once this is verified, we arrive at a contradiction with ① for which $k = \dfrac{y-1}{d}$.

If $k = 0$, the equation $yv^2 = (y+1)u^2$ has no positive integral solution (u, v). In the following, we assume $0 < k < y$. Suppose (u, v) is a positive integral solution of ②. For every integer m,

$$(u\sqrt{y+1}+v\sqrt{y})(\sqrt{y+1}-\sqrt{y})^{2m}(u\sqrt{y+1}-u\sqrt{y})(\sqrt{y+1}+\sqrt{y})^{2m} = -k.$$

The expansion of $(u\sqrt{y+1}+v\sqrt{y})(\sqrt{y+1}-\sqrt{y})^{2m}$ has the form

$$a\sqrt{y+1}+b\sqrt{y}, a, b \in \mathbb{N}$$

and its conjugate

$$(u\sqrt{y+1}-v\sqrt{y})(\sqrt{y+1}+\sqrt{y})^{2m} = a\sqrt{y+1}-b\sqrt{y}.$$

Therefore, $(y+1)a^2 - yb^2 = -k$, and (a, b) is another integral solution of ②. As $u\sqrt{y+1}+v\sqrt{y} > 1$ and $0 < \sqrt{y+1}-\sqrt{y} < 1$, we may choose integer m (unique) such that

$$\frac{1}{2} \le a\sqrt{y+1}+b\sqrt{y} < \frac{1}{2}(\sqrt{y+1}+\sqrt{y})^2. \qquad (*)$$

(1) If $a, b > 0$. Let $(s, t) = (a, b)$ be a positive integral solution of ② which satisfies

$$s\sqrt{y+1}+t\sqrt{y} < \frac{1}{2}(\sqrt{y+1}+\sqrt{y})^2. \qquad (**)$$

(2) If a, b are not both positive. From $yb^2 = (y+1)a^2 + k$, we have $|b|\sqrt{y} > |a|\sqrt{y+1}$, $|b| > |a|$. The former inequality of (*) implies $b > 0$. Hence $a > 0$, and

$$(-a)\sqrt{y+1} + b\sqrt{y} = \frac{k}{a\sqrt{y+1} + b\sqrt{y}} \leq 2k < 2y < \frac{1}{2}(\sqrt{y+1} + \sqrt{y})^2.$$

This implies that $(s, t) = (-a, b)$ is a positive integral solution of ② that satisfies (**).

From (**), we infer that

$$s\sqrt{y+1} < \frac{1}{2}(\sqrt{y+1} + \sqrt{y})^2 < 2(y+1), \text{ and thus } s < 2\sqrt{y+1} < 2y.$$

However,

$$k = yt^2 - (y+1)s^2 \geq y(s+1)^2 - (y+1)s^2 = y + s(2y - s) > y.$$

This is a contradiction. The proof is now complete. □

Test IV, First Day
(8:00 – 12:30; March 26, 2018)

1 Given a sequence of polynomials $\{f_n(x)\}_{n\geq 1}$, defined by:

$$f_1(x) = 1, f_{2n}(x) = xf_n(x), f_{2n+1}(x) = f_n(x) + f_{n+1}(x), n \geq 1.$$

Find $E = \{a \in \mathbb{Q} : f_n(a) = 0 \text{ for certain } n \geq 1\}$. (Contributed by Zhang Sihui)

Solution Since $f_2(x) = xf_1(x) = x$, $f_2(0) = 0$, $0 \in E$. In the following, we assume a is a root of $f_m(x)$, and $a \neq 0$. Let $m = 2^\alpha l$, $\alpha \geq 0$, l being odd. Notice that $f_m(x) = x^\alpha f_l(x)$, and $a \neq 0$. It follows $f_l(a) = 0$, $f_l(a)$ and $f_m(x)$ have the same nonzero roots, and so we only need to consider rational roots of $f_n(x)$ for odd n.

From the recurrence formula, we see that all the coefficients of $f_n(x)$ are nonnegative integers, the constant term of $f_{2n}(x)$ is 0, and the constant term of $f_{2n+1}(x)$ is 1. When n is odd, by the rational root theorem, all rational roots of $f_n(x)$ must be of the form $-\dfrac{1}{k}$, k being a positive integer. We assert $k < 4$.

Lemma   *If $k \geq 4$, then*

$$f_{2n+1}\left(-\frac{1}{k}\right) > \frac{1}{2}\max\left\{\left|f_n\left(-\frac{1}{k}\right)\right|,\left|f_{n+1}\left(-\frac{1}{k}\right)\right|\right\} > 0, n \geq 1. \quad \textcircled{1}$$

Proof of lemma   Use induction. If $n = 1$, $f_1\left(-\frac{1}{k}\right) = 1$, $f_2\left(-\frac{1}{k}\right) = -\frac{1}{k}$, $f_3\left(-\frac{1}{k}\right) = 1 - \frac{1}{k} > \frac{1}{2}$, $\textcircled{1}$ holds. Suppose that $n \geq 2$ and $\textcircled{1}$ holds for all $m < n$.

If $n = 2s$ is even,

$$f_{4s+1}\left(-\frac{1}{k}\right) = f_{2s}\left(-\frac{1}{k}\right) + f_{2s+1}\left(-\frac{1}{k}\right) = -\frac{1}{k}f_s\left(-\frac{1}{k}\right) + f_{2s+1}\left(-\frac{1}{k}\right)$$

$$\geq f_{2s+1}\left(-\frac{1}{k}\right) - \frac{1}{4}\left|f_s\left(-\frac{1}{k}\right)\right|$$

$$> f_{2s+1}\left(-\frac{1}{k}\right) - \frac{1}{2}f_{2s+1}\left(-\frac{1}{k}\right) = \frac{1}{2}f_{2s+1}\left(-\frac{1}{k}\right),$$

in which the last inequality comes from the induction hypothesis $f_{2s+1}\left(-\frac{1}{k}\right) > \frac{1}{2}\left|f_s\left(-\frac{1}{k}\right)\right|$. Again by induction, we have

$$f_{2s+1}\left(-\frac{1}{k}\right) > \frac{1}{2}\left|f_s\left(-\frac{1}{k}\right)\right| = \frac{k}{2}\left|f_{2s}\left(-\frac{1}{k}\right)\right| \geq \left|f_{2s}\left(-\frac{1}{k}\right)\right|.$$

Hence $\max\left\{\left|f_{2s}\left(-\frac{1}{k}\right)\right|,\left|f_{2s+1}\left(-\frac{1}{k}\right)\right|\right\} = f_{2s+1}\left(-\frac{1}{k}\right) > 0$, and

$$f_{4s+1}\left(-\frac{1}{k}\right) > \frac{1}{2}f_{2s+1}\left(-\frac{1}{k}\right) = \frac{1}{2}\max\left\{\left|f_{2s}\left(-\frac{1}{k}\right)\right|,\left|f_{2s+1}\left(-\frac{1}{k}\right)\right|\right\} > 0.$$

If $n = 2s + 1$ is odd,

$$f_{4s+3}\left(-\frac{1}{k}\right) = f_{2s+1}\left(-\frac{1}{k}\right) + f_{2s+2}\left(-\frac{1}{k}\right)$$

$$= f_{2s+1}\left(-\frac{1}{k}\right) - \frac{1}{k}f_{s+1}\left(-\frac{1}{k}\right)$$

$$\geq f_{2s+1}\left(-\frac{1}{k}\right) - \frac{1}{4}\left|f_{s+1}\left(-\frac{1}{k}\right)\right|$$

$$> f_{2s+1}\left(-\frac{1}{k}\right) - \frac{1}{2}f_{2s+1}\left(-\frac{1}{k}\right) = \frac{1}{2}f_{2s+1}\left(-\frac{1}{k}\right),$$

in which the last inequality comes from the induction hypothesis $f_{2s+1}\left(-\frac{1}{k}\right) > \frac{1}{2}\left|f_{s+1}\left(-\frac{1}{k}\right)\right|$. Again by induction, we have

$$f_{2s+1}\left(-\frac{1}{k}\right) > \frac{1}{2}\left|f_{s+1}\left(-\frac{1}{k}\right)\right| = \frac{k}{2}\left|f_{2s+2}\left(-\frac{1}{k}\right)\right| \geq \left|f_{2s+2}\left(-\frac{1}{k}\right)\right|.$$

Hence $\max\{\left|f_{2s+1}\left(-\frac{1}{k}\right)\right|, \left|f_{2s+2}\left(-\frac{1}{k}\right)\right|\} = f_{2s+1}\left(-\frac{1}{k}\right) > 0$, and

$$f_{4s+3}\left(-\frac{1}{k}\right) > \frac{1}{2}f_{2s+1}\left(-\frac{1}{k}\right) = \frac{1}{2}\max\left\{\left|f_{2s+1}\left(-\frac{1}{k}\right)\right|,\right.$$

$$\left.\left|f_{2s+2}\left(-\frac{1}{k}\right)\right|\right\} > 0.$$

This completes the induction. From (1), we deduce that if $k \geq 4$, $-\frac{1}{k}$ is not a root of any $f_n(x)$.

On the other hand, $f_3(x) = x + 1$, $f_5(x) = 2x + 1$, $f_{21}(x) = 3x^2 + 4x + 1$, they have roots $-1, -\frac{1}{2}$, and $-\frac{1}{3}$, respectively. Therefore $E = \{0, -1, -\frac{1}{2}, -\frac{1}{3}\}$. □

2 Among 32 students in a class, there are 10 social groups, each with exactly 16 students. For every two students, call the square of the number of groups that exactly one of them belong to as their "social disparity". Let S be the sum of all $C_{32}^2 (= 496)$ social disparities. Find the minimum possible value of S. (Contributed by Wang Bin)

Solution Let $n = 32$, $m = 10$. We use an $n \times m$ table $\{x_{i,k}\}$ to record the membership information: if the i-th student belongs to the k-th social group, then $x_{i,k} = +1$; otherwise, $x_{i,k} = -1$, in which $i = 1, 2, \ldots n, k = 1, 2, \ldots, m$.

Let $d_{i,j}$ be the number of groups that exactly one of the i-th student and the j-th student attend. Evidently, their social disparity is $d_{i,j}^2$, and thereby S is the summation of all $d_{i,j}^2$'s, $i<j$. We have $2S = \sum_{i=1}^{n}\sum_{j=1}^{n} d_{i,j}^2$ (as $d_{i,i} = 0$).

Let $r_i = (x_{i,1}, x_{i,2}, \ldots, x_{i,m}), i = 1, 2, \ldots, n$, be the i-th row vector (it includes all information of the i-th student); let

$$c_k = (x_{1,k}, x_{2,k}, \ldots, x_{n,k})^T, k = 1, 2, \ldots, m,$$

be the k-th column vector (it includes all information of the k-th group). Since every social group contains $16 = \frac{n}{2}$ students, c_k has equal numbers of $+1$'s and -1's, and hence $\sum\limits_{i=1}^{n} x_{i,k} = 0$, i.e., the sum of all entries of a column is zero.

Define the inner product of the k-th and the l-th column vectors as $C_{k,l} = \sum\limits_{i=1}^{n} x_{i,k} x_{i,l}$, the inner product of the i-th and the j-th row vectors as $R_{i,j} = \sum\limits_{k=1}^{m} x_{i,k} x_{j,k}$ (*). Since row i and row j differ in $d_{i,j}$ columns, we infer in (*) $d_{i,j}$ of the summands are -1's and the other $(m - d_{i,j})$ of them are $+1$'s, and so $R_{i,j} = m - 2d_{i,j}$.

Consider $T_1 = \sum\limits_{i=1}^{n} \sum\limits_{j=1}^{n} R_{i,j}$, the sum of all n^2 row inner products:

$$T_1 = \sum_{i=1}^{n} \sum_{j=1}^{n} \sum_{k=1}^{m} x_{i,k} x_{j,k} = \sum_{k=1}^{m} \left(\sum_{i=1}^{n} \sum_{j=1}^{n} x_{i,k} x_{j,k} \right)$$

$$= \sum_{k=1}^{m} \left(\sum_{i=1}^{n} x_{i,k} \right) \left(\sum_{j=1}^{n} x_{j,k} \right) = 0.$$

The sum of all $d_{i,j}$'s is a fixed number $\dfrac{mn^2}{2}$, and this implies the average value of $d_{i,j}$'s is $\dfrac{m}{2}$. To minimize $2S = \sum\limits_{i,j=1}^{n} d_{i,j}^2$, it is equivalent to minimizing the variance $\sum\limits_{i,j=1}^{n} \left(d_{i,j} - \dfrac{m}{2} \right)^2$, or minimizing $T_2 = \sum\limits_{i,j=1}^{n} R_{i,j}^2$. Notice that

$$T_2 = \sum_{i=1}^{n} \sum_{j=1}^{n} R_{i,j}^2 = \sum_{i=1}^{n} \sum_{j=1}^{n} \left(\sum_{k=1}^{m} x_{i,k} x_{j,k} \right)^2$$

$$= \sum_{i=1}^{n} \sum_{j=1}^{n} \left(\sum_{k=1}^{m} \sum_{l=1}^{m} x_{i,k} x_{j,k} x_{i,l} x_{j,l} \right)$$

$$= \sum_{k=1}^{m} \sum_{l=1}^{m} \left(\sum_{i=1}^{n} \sum_{j=1}^{n} x_{i,k} x_{i,l} x_{j,k} x_{j,l} \right)$$

$$= \sum_{k=1}^{m} \sum_{l=1}^{m} \left(\sum_{i=1}^{n} x_{i,k} x_{i,l} \right)^2 = \sum_{k=1}^{m} \sum_{l=1}^{m} (C_{k,l})^2.$$

Hence the sum of the squares of all the row inner products is equal to the sum of the squares of all the column inner products.

Since $C_{k,k} = n, T_2 = \sum\limits_{k,l=1}^{m} (C_{k,l})^2 \geq \sum\limits_{k=1}^{m} (C_{k,k})^2 = mn^2$. It follows that

$$2S = \sum_{i,j=1}^{n} d_{i,j}^2 = \sum_{i,j=1}^{n} \left[\left(\frac{m - 2d_{i,j}}{2} \right)^2 + \left(\frac{m}{2} \right)^2 - \frac{m}{2}(m - 2d_{i,j}) \right]$$

$$= \sum_{i,j=1}^{n} \left(\frac{R_{i,j}}{2} \right)^2 + n^2 \times \frac{m^2}{4} - \frac{m}{2} \sum_{i,j=1}^{n} R_{i,j}$$

$$= \frac{1}{4} T_2 + \frac{n^2 m^2}{4} - \frac{m}{2} T_1$$

$$\geq \frac{mn^2}{4} + \frac{n^2 m^2}{4} = \frac{n^2 m(m+1)}{4},$$

and we conclude $S \geq \dfrac{n^2 m(m + 1)}{8} = 14080$.

The above minimum may be attained. Construct the 32×32 Hadamard matrix H_{32}, for which the first column is all $+1$'s. Take column 2 through column 11 of H_{32} as the 32 by 10 table $\{x_{i,k}\}$. Since all the column vectors are pairwise orthogonal, $C_{k,l} = 0$ for $k \neq l$. We have $T_2 = mn^2$ and $S = 14080$. $\qquad\square$

Remark Denote $\{x_{i,k}\}$ by an $n \times m$ matrix A. Then the n^2 row inner products correspond to the n^2 entries of the symmetric matrix $B_1 = AA^T$, and the sum of the squares of them is equal to the trace of $B_1 B_1^T = B_1^2$, which is also equal to the sum of the squares of the eigenvalues of B_1. Meanwhile, the sum of the squares of the m^2 column inner products is equal to the sum of the squares of the eigenvalues of the symmetric matrix $B_2 = A^T A$. When $n \geq m$, the eigenvalues of $B_1 = AA^T$ (n of them) are exactly those of $B_2 = A^T A$ (m of them) plus $(n - m)$ of 0's. Therefore the sum of the squares are equal.

We may apply the singular value decomposition to A: let $A = U\Sigma V$, where U is an $n \times n$ orthogonal matrix, V is an $m \times m$ orthogonal matrix, and $\Sigma = \{\sigma_{i,k}\}$ is an $n \times m$ rectangular diagonal matrix, i.e., all entries other than $\sigma_{1,1}, \sigma_{2,2}, \ldots, \sigma_{m,m}$ vanish. Now $B_1 = AA^T = (U\Sigma V)(V^T \Sigma^T U^T) = U(\Sigma\Sigma^T)U^T$, where $\Sigma\Sigma^T$ is an $n \times n$ diagonal matrix with diagonal entries $\sigma_{1,1}^2, \sigma_{2,2}^2, \ldots, \sigma_{m,m}^2, 0, \ldots, 0$, and B_1 is similar to $\Sigma\Sigma^T$. The eigenvalues of $B_1 = AA^T$ are (n of them)

$$\sigma_{1,1}^2, \sigma_{2,2}^2, \ldots, \sigma_{m,m}^2, 0, \ldots, 0.$$

Analogously, the eigenvalues of $B_2 = A^T A$ are (m of them)

$$\sigma_{1,1}^2, \sigma_{2,2}^2, \ldots, \sigma_{m,m}^2.$$

For $A = U\Sigma V$, the nonzero entries of Σ (all among the diagonal entries $\sigma_{1,1}, \sigma_{2,2}, \ldots, \sigma_{m,m}$) are called the singular values of A; A and A^T have the same singular values. In the solution, T_2 represents the sum of the fourth powers of all singular values of A.

3 As shown in Fig. 3.1, in isosceles $\triangle ABC, AB = AC$. Let points D, E, F lie on the sides BC, AC, AB, respectively, such that $DE//AB, DF//AC$. Let ω_1, the circumcircle of $\triangle ABC$, and ω_2, the circle passing through A, E, F, intersect at A and G. Let DE and ω_2 meet at E and K; points L, M are on ω_1, ω_2, respectively, such that $LG \perp KG$, $MG \perp CG$. Denote P as the circumcenter of $\triangle DGL$, Q as the circumcenter of $\triangle DGM$, as shown in figure ①. Prove A, G, P, Q are concyclic. (Contributed by He Yijie)

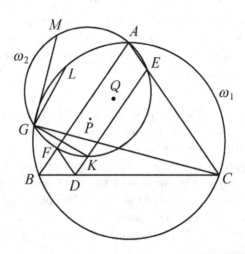

Fig. 3.1

Solution As shown in Fig. 3.2, let O be the center of ω_1. Evidently, we have $\triangle AOB \cong \triangle COA$. Since $BF = DF = AE$, when $\triangle AOB$ is rotated around O to $\triangle COA$, $\triangle BOD$ becomes $\triangle AOE$, and thus

$$\angle EOF = \angle AOB = \pi - \angle EAF, O \text{ is on } \omega_2.$$ It follows from $EK//AF$ that $FK = AE = FD = FB$; from G, K, E, A concyclic that

$$\angle GKE = \pi - \angle GAE = \angle GBC.$$

Hence B, D, K, G lie on a circle centered at $F, FG = FD = AE$, which further implies $EF//AG$, and thus $EG = AF = ED$. Moreover,

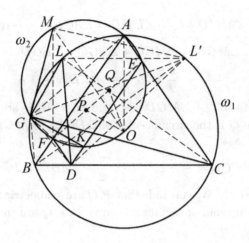

Fig. 3.2

$PQ = PD, QG = QD$, and we deduce that E, F, P, Q are all on the perpendicular bisector of DG.

Let l be the perpendicular bisector of AG. As $OA = OG, O$ lies on l; as $EF//AG, E, F$ are symmetric about l. We assert $\angle EPO = \angle FQO$, which implies P, Q are also symmetric about l.

Denote $\angle BAC = 2\alpha$. Since P is the circumcenter of $\triangle DGL, OP \perp GL$; from $GL \perp GK$, it follows $OP//GK$. Then from $EF//AG$ we deduce

$$\angle EPO = \angle AGK = \angle AFK = \angle EAF = 2\alpha.$$

Now we turn to $\angle FQO$. Notice that $CL \perp AB$, because

$$\angle LCB + \angle ABC = (\pi - \angle LGB) + \angle ABC$$
$$= \pi - \left(\frac{\pi}{2} + \angle KGB\right) + \angle EDC = \frac{\pi}{2}.$$

Let $LL'//AB, L'$ being the intersection with ω_1. Likewise, $BL' \perp AC$. We show that M is the circumcenter of $\triangle BDL'$. Since $\angle AGM = \frac{\pi}{2} - \angle AGC = \frac{\pi}{2} - \angle ABC = \alpha$, on ω_2 we have $\widehat{AM} = \widehat{OF} = \widehat{OE}$, thus $FM \parallel AO, AE \parallel OM$, and further $FM \perp BC, OM \perp BL'$. From $FB = FD$ and $OB = OL'$, we infer that line FM is the perpendicular bisector of BD, and line OM is the perpendicular bisector of BL'. Hence M is the circumcenter of $\triangle BDL'$, and $\angle ML'D = \dfrac{\pi}{2} - \angle L'BD = \angle ACB = \dfrac{\pi}{2} - \alpha$. Notice that

$$\angle MGD = \angle MGC + \angle BGC - \angle BGD$$

$$= \frac{\pi}{2} + 2\alpha - \frac{1}{2}\angle BFD = \frac{\pi}{2} + \alpha.$$

It follows that $\angle ML'D + \angle MGD = \pi$, and D, G, M, L' are concyclic.

Finally, since Q is the circumcenter of $\triangle DGL'$, $OQ \perp GL'$. From $EF \parallel AG$, we obtain

$$\angle FQO = \frac{\pi}{2} - \angle AGL' = \frac{\pi}{2} - \angle ABL' = 2\alpha.$$

So $\angle EPO = \angle FQO$. We conclude that P, Q are symmetric about l, $AGPQ$ is an isosceles trapezoid or rectangle, and A, G, P, Q are concyclic. $\qquad\square$

Test IV, Second Day
(8:00 – 12:30; March 27, 2018)

4 Let p be a prime and k be a positive integer. Set S consists of all integers a satisfying: $1 \le a \le p-1$, and there exists positive integer x, such that $x^k \equiv a \pmod{p}$. Suppose that $3 \le |S| \le p-2$. Prove: the elements of S do not form an arithmetic sequence. (Contributed by Yu Hongbing)

Solution Assume that the elements of $S, a_1 < a_2 < \cdots < a_n$, form an arithmetic sequence. We shall try to derive a contradiction. As $n \ge 3$, there exists $a \in S$, such that $a^2 \not\equiv 1 \pmod{p}$. By the given condition, there exists positive integer x, such that $x^k \equiv a \pmod{p}$; and for any $b \in S$, there exists positive integer y, such that $y^k \equiv b \pmod{p}$. Then $(xy)^k \equiv x^k y^k \equiv ab \pmod{p}$, i.e., $ab \, x \in S$. Therefore, $aa_1, aa_2, \ldots, aa_k \pmod{p}$ is a permutation of $a_1, a_2, \ldots, a_k$. It follows that

$$\sum_{i=1}^{n} aa_i \equiv \sum_{i=1}^{n} a_i \pmod{p} \Rightarrow \sum_{i=1}^{n} a_i \equiv 0 \pmod{p}.$$

$$\sum_{i=1}^{n} (aa_i)^2 \equiv \sum_{i=1}^{n} a_i^2 \pmod{p} \Rightarrow \sum_{i=1}^{n} a_i^2 \equiv 0 \pmod{p}.$$

Let $d = a_2 - a_1$ be the common difference, $a_i = a_1 + (i-1)d, i = 1, 2, \ldots, n$. From

$$\sum_{i=1}^{n} a_i = \sum_{i=1}^{n} (a_1 + (i-1)d) = na_1 + \frac{1}{2}n(n-1)d \equiv 0 \pmod{p},$$

we have

$$2a_1 + (n-1)d \equiv 0 \pmod{p}. \qquad \text{(1)}$$

From

$$\sum_{i=1}^{n} a_i^2 = \sum_{i=1}^{n} (a_1 + (i-1)d)^2 = na_1^2 + n(n-1)a_1 d$$
$$+ \frac{1}{6}(n-1)n(2n-1)d^2 \equiv 0 \pmod{p},$$

multiply by 6 on both sides and cancel n; then use $(n-1)d \equiv -2a_1 \pmod{p}$ from ① to derive

$$6a_1^2 + 6a_1(-2a_1) + (2n-1)(-2a_1)d = -6a_1^2 - 2(2n-1)a_1 d \equiv 0 \pmod{p}.$$

We simplify the above equation, obtaining

$$6a_1 + 2(2n-1)d \equiv 0 \pmod{p}. \qquad \text{(2)}$$

It follows from ① and ② that

$$(2(2n-1) - 3(n-1))d = (n+1)d \equiv 0 \pmod{p}.$$

However, $3 \le n < p-1, 0 < d < p$, and the above equation cannot hold. This contradiction indicates that the elements of S cannot form an arithmetic sequence. $\square$

5 Given $\lambda \in (0,1)$ and positive integer n, prove that the modulus of the roots of polynomial

$$f(x) = \sum_{k=0}^{n} C_n^k \lambda^{k(n-k)} x^k$$

are all equal to 1. (Contributed by Yao Yijun)

Solution Let $f_n(x) = \sum_{k=0}^{n} C_n^k \lambda^{k(n-k)} x^k$. To prove the statement, we apply induction on n. When $n = 1$, $f_1(x) = x + 1$, and the conclusion holds. Assume it holds for n, and the roots are $z_1, z_2, \ldots, z_n, |z_j| = 1$. Since $f_n(x)$

is monic,

$$f_n(x) = \prod_{j=1}^{n}(x - z_i).$$

By definition,

$$f_{n+1}(x) = \sum_{k=0}^{n+1} C_{n+1}^{k} \lambda^{k(n+1-k)} x^k$$

$$= \sum_{k=1}^{n}(C_n^{k-1} + C_n^{k})\lambda^{k(n+1-k)} x^k + 1 + x^{n+1}$$

$$= \sum_{k=0}^{n} C_n^{k}\lambda^{(k+1)(n-k)} x^{k+1} + \sum_{k=0}^{n} C_n^{k}\lambda^{k(n+1-k)} x^k$$

$$= x\lambda^n f_n\left(\frac{x}{\lambda}\right) + f_n(\lambda x).$$

If z is a root of $f_{n+1}(x)$, clearly $z \neq 0$, and from the above identity, we obtain

$$z\lambda^n \prod_{j=1}^{n}\left(\frac{z}{\lambda} - z_j\right) + \prod_{j=1}^{n}(\lambda z - z_j) = 0,$$

which implies

$$z\prod_{j=1}^{n}(z - \lambda z_j) = -\prod_{j=1}^{n}(\lambda z - z_j). \qquad \text{①}$$

From $|z_j| = 1$, direct computation yields

$$|z - \lambda z_j|^2 - |\lambda z - z_j|^2 = (z - \lambda z_j)(\bar{z} - \lambda\bar{z_j}) - (\lambda z - z_j)(\lambda\bar{z} - \bar{z_j})$$

$$= (1 - \lambda^2)(|z|^2 - 1).$$

If $|z| > 1, |z - \lambda z_j| > |\lambda z - z_j|, j = 1, 2, \ldots, n$, taking modulus in ① implies $|z| < 1$, a contradiction. If $|z| < 1, |z - \lambda z_j| < |\lambda z - z_j|, j = 1, 2, \ldots, n$, taking modulus in ① implies $|z| > 1$, a contradiction again. So it must be $|z| = 1$. □

6 Let $a_i, b_i, c_i(i = 1, 2, \ldots, n)$ be $3n$ reals in the interval $[0, 1]$. Define

$$S = \{(i, j, k) \mid a_i + b_j + c_k < 1\}, T = \{(i, j, k) \mid a_i + b_j + c_k > 2\}.$$

Given that $|S| \geq 2018$ and $|T| \geq 2018$, find the smallest possible value of n. (Contributed by Fu Yunhao)

Solution The smallest possible n is 20.

First, take $a_i = b_i = c_i = 0.3, i = 1, 2, \ldots, 13; a_j = b_j = c_j = 0.9, j = 14, 15, \ldots, 20$. Then $S = \{(i, j, k) \mid 1 \le i, j, k \le 13\}, |S| = 13^3 = 2197$, while

$$T = \{(i, j, k) \mid 1 \le i, j, k \le 20, \text{ at least two of them are } \ge 14\},$$

$|T| = 7^3 + 3 \times 7^2 \times 13 = 2254$. The conditions are satisfied.

Now we prove $n \ge 20$. In fact, we claim a stronger result as follows. Let l, m, n be positive integers, $a_1, \ldots, a_l, b_1, \ldots, b_m, c_1, \ldots, c_n$ are reals in the interval $[0, 1]$, and S, T are defined as in the original problem. Then

$$\min\{|S|, |T|\} \le Clmn, \qquad \text{①}$$

where $C = \left(2\cos\dfrac{2\pi}{9}\right)^{-3} \in [0.27, 0.28]$.

Assume that there exist $l, m, n, a_1, \ldots, a_l, b_1, \ldots, b_m, c_1, \ldots, c_n$, such that

$$|S| > Clmn, |T| > Clmn.$$

We shall arrive at a contradiction at the end. Denote

$$S_k = \{(i, j) \mid (i, j, k) \in S\},$$

$$T_k = \{(i, j) \mid (i, j, k) \in T\}, 1 \le k \le n.$$

Evidently, $|S| = \sum\limits_{k=1}^{n} |S_k|, |T| = \sum\limits_{k=1}^{n} |T_k|$, and

$$\frac{|S|}{lmn} = \frac{1}{n} \sum_{k=1}^{n} \frac{|S_k|}{lm} > C,$$

$$\frac{|T|}{lmn} = \frac{1}{n} \sum_{k=1}^{n} \frac{|T_k|}{lm} > C.$$

Let $v_k = \left(\dfrac{|S_k|}{lm}, \dfrac{|T_k|}{lm}\right) \in \mathbb{R}^2, 1 \le k \le n$. The above inequalities indicate that both coordinates of the arithmetic mean of $v_1, v_2, \ldots, v_n$ are greater than C. We need a lemma as follows.

Lemma *There exist $1 \le k_1 < k_2 \le n$ and rational number $\lambda \in [0, 1]$, such that both coordinates of $\lambda v_{k_1} + (1 - \lambda)v_{k_2}$ are greater than C.*

Proof of lemma Let P be the convex hull of $v_1, v_2, \ldots, v_n$. Clearly, $v \in P$. Since

$$v \in \Omega = \{(x,y) \in \mathbb{R}^2 \,|\, x > C, y > C\},$$

we deduce $\Omega \cap \partial P \neq \emptyset$, where ∂P is the boundary of P. Hence there exist $k_1 \neq k_2$, such that the segment

$$\{\lambda v_{k_1} + (1 - \lambda)v_{k_2} \,|\, 0 \le \lambda \le 1\}$$

intersects with Ω. Since Ω is open and $\mathbb{Q}$ is dense in $\mathbb{R}$, there exists a rational number $\lambda \in [0, 1]$, such that $\lambda v_{k_1} + (1 - \lambda)v_{k_2} \in \Omega$. The proof of lemma is complete.

Let $\lambda = \dfrac{p}{q}, q > 0, 0 \le p \le q$. Consider $(l + m + q)$ reals $a_1', \ldots, a_l', b_1', \ldots, b_m', c_1', \ldots, c_q' \in [0, 1]$, in which $a_i' = a_i, 1 \le i \le l, b_j' = b_j, 1 \le j \le m, c_1' = \cdots = c_p' = c_{k_1}, c_{p+1}' = \cdots = c_q' = c_{k_2}$. Define S' and T' accordingly. Then

$$S' = \{(i,j,k) \,|\, (i,j) \in S_{k_1}, 1 \le k \le p\} \cup$$

$$\{(i,j,k) \,|\, (i,j) \in S_{k_2}, p+1 \le k \le q\},$$

$$T' = \{(i,j,k) \,|\, (i,j) \in T_{k_1}, 1 \le k \le p\} \cup$$

$$\{(i,j,k) \,|\, (i,j) \in T_{k_2}, p+1 \le k \le q\},$$

and we have

$$\frac{|S'|}{lmp} = \frac{q|S_{k_1}|}{lmp} + \frac{(p-q)|S_{k_2}|}{lmp} = \lambda \frac{|S_{k_1}|}{lm} + (1-\lambda)\frac{|S_{k_2}|}{lm} > C,$$

$$\frac{|T'|}{lmp} = \frac{q|T_{k_1}|}{lmp} + \frac{(p-q)|T_{k_2}|}{lmp} = \lambda \frac{|T_{k_1}|}{lm} + (1-\lambda)\frac{|T_{k_2}|}{lm} > C.$$

From the above argument, we realize that if $(l + m + q)$ reals have $|S|, |T| > Clmq$, then by modifying the values of $c_1, c_2, \ldots, c_q$, the corresponding index sets S, T may still have more than $Clmq$ elements, but $c_1, c_2, \ldots, c_n$ take at most two distinct values. Similarly, we can modify $a_1, \ldots, a_l$ and $b_1, \ldots, b_m$, such that all a_i's take at most two distinct values, all b_j's take at most two distinct values, and $|S|, |T| > Clmq$. So essentially, we may focus on this special situation:

$a_1, \ldots, a_l, b_1, \ldots, b_m, c_1, \ldots, c_n$, satisfy that $a_1, \ldots, a_l$ take only $a_1 \le a_2$ (equality means taking only one value), $b_1, \ldots, b_m$ take only $b_1 \le b_2, c_1, \ldots, c_n$ take only $c_1 \le c_2$; and $|S|, |T| > Clmq$.

Let x of a_i's equal a_1, y of b_j's equal b_1, and z of c_k's equal c_1. Then $a_i + b_j + c_k$ attains at most 8 values, $\{a_1, a_2\} + \{b_1, b_2\} + \{c_1, c_2\}$. Clearly

$a_1 + b_1 + c_1$ is the smallest and $a_2 + b_2 + c_2$ is the largest. From $|S|, |T| \neq 0$, we have $a_1 + b_1 + c_1 < 1, a_2 + b_2 + c_2 > 2$. There are three cases.

Case 1: All the other (≤ 6) values are in $[1, 2]$. Then

$$|S| = xyz, |T| = (l-x)(m-y)(n-z).$$

Due to

$$\frac{x}{l} + \frac{y}{m} + \frac{z}{n} + \frac{l-x}{l} + \frac{m-y}{m} + \frac{n-z}{n} = 3,$$

if $\dfrac{x}{l} + \dfrac{y}{m} + \dfrac{z}{n} \leq \dfrac{3}{2}$, by the AM-GM inequality

$$\frac{|S|}{lmn} = \frac{xyz}{lmn} \leq \frac{1}{8} < C,$$

a contradiction; otherwise $|T| < Clmn$, again, a contradiction.

In the following, certain value(s) are not in $[1, 2]$. By symmetry, we may assume $a_1 + b_1 + c_2 \notin [1, 2]$. Then from $a_1 + b_1 + c_2 \leq a_1 + b_1 + c_1 + 1 \leq 2$, it follows $a_1 + b_1 + c_2 < 1$, and likewise, $a_1 + b_2 + c_1, a_2 + b_1 + c_1, a_1 + b_2 + c_2, a_2 + b_1 + c_2$ are all less than 2. There are two cases from here.

Case 2: $a_2 + b_2 + c_1 > 2$. Then $a_1 + b_2 + c_1, a_2 + b_1 + c_1, a_1 + b_2 + c_2, a_2 + b_1 + c_2$ are all greater than 1, and thus

$$|S| = xyz + xy(n-z) = xyn,$$

$$|T| = (l-x)(m-y)(n-z) + (l-x)(m-y)z = (l-x)(m-y)n.$$

From $\dfrac{|S|}{lmn} = \dfrac{xy}{lm} > C > \dfrac{1}{4}$, we get $\dfrac{x}{l} + \dfrac{y}{m} > 1$, and likewise $\dfrac{l-x}{l} + \dfrac{m-y}{m} > 1$. However,

$$2 = \left(\frac{x}{l} + \frac{y}{m}\right) + \left(\frac{l-x}{x} + \frac{m-y}{m}\right) > 2,$$

a contradiction.

Case 3: $a_2 + b_2 + c_1 \leq 2$. Then $|T| = (l-x)(m-y)(n-z)$. Notice that $a_1 + b_2 + c_2, a_2 + b_1 + c_2, a_2 + b_2 + c_1$ are greater than 1; at most

$a_1 + b_1 + c_2, a_1 + b_2 + c_1, a_2 + b_1 + c_1$ can be less than 1, and thus

$$|S| \leq xyz + xy(n - z) + xz(m - y) + yz(l - x). \tag{*}$$

Let $r = \dfrac{l-x}{l}$, $s = \dfrac{m-y}{m}$, $t = \dfrac{n-z}{n}$, $r, s, t \in [0, 1]$. We must have

$$rst = \frac{|T|}{lmn} > C, \tag{**}$$

and (*) becomes

$$(1-r)(1-s)(1-t) + (1-r)(1-s)t + (1-r)s(1-t) + r(1-s)(1-t) \geq \frac{|S|}{lmn} > C,$$

which can be simplified to

$$1 - (rs + st + tr) + 2rst > C.$$

Let $rst = d^3$. By the AM-GM inequality, $rs + st + tr \geq 3d^2$, and thus

$$1 - 3d^2 + 2d^3 \geq 1 - (rs + st + tr) + 2rst > C.$$

From (**), $d > \sqrt[3]{C} = \left(2\cos\dfrac{2\pi}{9}\right)^{-1}$. Since $1 - 3x^2 + 2x^3$ is decreasing on $[0, 1]$,

$$C < 1 - 3d^2 + 2d^3 < 1 - 3C^{\frac{2}{3}} + 2C$$

$$= C\left(\left(2\cos\frac{2\pi}{9}\right)^3 - 3 \cdot 2\cos\frac{2\pi}{9} + 2\right)$$

$$= C\left(2\cos\frac{6\pi}{9} + 2\right) = C,$$

and we arrive at a contradiction. This validates ①. In the original problem, if $n \leq 19$,

$$\min\{|S|, |T|\} \leq Cn^3 \leq 19^3 \times 0.28 = 1920.52 < 2018,$$

contradicting the assumption. Therefore $n \geq 20$.   □

China Girls' Mathematical Olympiad

2016 (Beijing)

The 15th Chinese Girls' Mathematical Olympiad (CGMO) was held in Beijing No. 4 Middle School from August 10th to 15th, 2016. A total of 152 female students from 38 national and regional teams participated in the game, including those from China's most provinces, municipalities, autonomous regions, Hong Kong Special Administrative Region, Macau Special Administrative Region, as well as from Russia, Philippines, Singapore, etc. After two games (4 hours for 4 questions each). 27 students headed by Zhao Lanxin won the gold medal (first prize), 60 students headed by Ye Wenqin won the silver medal (second prize), and 60 students headed by He Yaxin won the bronze medal (third prize). The top 13 gold medal winners from mainland China were invited to participate in the National Middle School Mathematics Winter Camp held in November 2016.

In order to enrich the lives of the contestants, the CGMO Organizing Committee also arranged women's aerobics competitions, gala evenings and visits to the National Museum of China, which created a strong cultural atmosphere for the mathematical competition. The contestants showed their skills through these colorful activities, to promote innovation, strengthen cooperation and increase exchanges and friendship.

Director of the Main Test Committee: Li Shenghong.

Members of the Main Test Committee (in alphabetical order):

Ai Yinghua (Tsinghua University);

Fu Yunhao (Guangdong Second Normal University);

Ji Chungang (Nanjing Normal University);

Li Shenghong (Zhejiang University);

Li Wei Gu (Peking University);

Liang Yingde (University of Macau);

Luo Wei (Zhejiang University);

Wang Bin (Institute of Mathematics and Systems Science, Chinese Academy of Sciences);

Wang Xinmao (University of Science and Technology of China).

First Day
(8:00 – 12:00; August 12, 2016)

1 Let integer $n \geq 3$. Put n^2 cards labeled $1, 2, \ldots, n^2$ into n boxes, each with n cards. The following operation is allowed: select two boxes, take two cards from each of these boxes, and put them into the other box. Prove: no matter how you place them initially, you can always go through a limited number of operations so that the card numbers in each box are consecutive n integers.

Solution We first prove that it is possible to exchange one card for any two boxes with two operations, and leave the positions of the remaining cards unchanged.

Suppose one of the given two boxes has three cards a, x, and y (the remaining cards are not considered for the time being), and the other box has three cards of b, z, and w (the remaining cards are not considered for the time being), where a and b are to be exchanged. The first operation exchanges a, x with z, w, and the second operation exchanges z, w with b, x, so that the three cards in the first box become b, x, y, and those in the second box become a, z, w. In this way, you can place the cards labeled $1, 2, \ldots, n$ into the first box, and then place the cards labeled $n + 1, n + 2, \ldots, 2n$ into the second box, with the cards in the first box remain unchanged. And so on, untill the cards labeled $n^2 - 2n + 1, n^2 - 2n + 2, \ldots, n^2 - n$ are placed into the $(n-1)$th box, with the cards in the previous boxes remain unchanged. At this time, the cards labeled $n^2 - n + 1, n^2 - n + 2, \ldots, n^2$ must be in the last box. The proof is then completed.   □

2 Suppose the lengths of the three sides of $\triangle ABC$ are $BC = a$, $AC = b$, $AB = c$, and Γ is the circumcircle of $\triangle ABC$.

(1) If there is a unique point $P(P \neq B, P \neq C)$ on $\overset{\frown}{BC}$ (not containing A) of Γ, satisfying $PA = PB + PC$, find the necessary and sufficient conditions for a, b, and c.

(2) Let P be the only point described in (1). Prove: if AP bisects segment BC, then $\angle BAC < 60°$.

Solution (1) If the given condition holds, according to Ptolemy's theorem, we have

$$a \cdot PA = b \cdot PB + c \cdot PC.$$

Combing it with $PA = PB + PC$, we get

$$(b - a)PB + (c - a)PC = 0.$$

If $b = a$, then obviously $c = a$. At this time, $\triangle ABC$ is an equilateral triangle. Then according to Ptolemy's theorem, any point Q on $\overparen{BC}$ (not containing A) of Γ satisfies $QA = QB + QC$, contradicting the uniqueness of P. Therefore $b \neq a$, and by the same reason $c \neq a$. So it is easy to see from the equation that either $b < a < c$ or $c < a < b$.

On the other hand, if either $b < a < c$ or $c < a < b$, then by Ptolemy's theorem we knows that, for any point P on $\overparen{BC}$, $PA = PB + PC$ is equivalent to

$$(b - a)PB + (c - a)PC = 0.$$

i.e., $\dfrac{PB}{PC} = \dfrac{c - a}{a - b} > 0$. Let AP intersects BC at K. Then

$$\frac{BK}{CK} = \frac{S_{\triangle ABP}}{S_{\triangle ACP}} = \frac{AB \cdot BP}{AC \cdot CP} = \frac{c}{b} \cdot \frac{PB}{PC}.$$

Therefore, $\dfrac{PB}{PC} = \dfrac{c - a}{a - b}$ is equivalent to $\dfrac{BK}{CK} = \dfrac{c(c - a)}{b(a - b)}$. As $\dfrac{c(c - a)}{b(a - b)} > 0$, then point K exists uniquely.

In summary, the necessary and sufficient conditions required are $b < a < c$ or $c < a < b$.

(2) By the given condition, we know $BK = CK$. Combining it with the result in (1), we get $c(c - a) = b(a - b)$, or equivalently.

$$a(b + c) = b^2 + c^2.$$

Then,

$$\cos \angle BAC = \frac{b^2 + c^2 - a^2}{2bc} = \frac{ab + ac - a^2}{2bc}$$

$$= \frac{1}{2} + \frac{(b - a)(a - c)}{2bc} > \frac{1}{2}.$$

Therefore, $\angle BAC < 60°$. The proof is completed. $\square$

3 Let m and n be coprime integers both greater than 1. Prove that there are positive integers a, b, and c that satisfy

$$m^a = 1 + n^b c, \text{ and } c \text{ is coprime with } n.$$

Solution (1) For any non-zero integer t and prime number p, we define $v_p(t)$ as the power of p contained in t, that is, the largest non-negative integer α satisfying $p^\alpha \mid t$. We first prove the following lemma.

Lemma *If d is an integer greater than 1, s is a positive integer, p is a prime number, and $v_p(d-1) = u \geq 2$, then*

$$v_p(d^s - 1) = u + v_p(s).$$

Proof of the lemma Let $e = d - 1$. Then $d^s - 1 = (e+1)^s - 1 = \sum_{i=1}^{S} e^i \cdot C_s^i$.

For $i = 1$, $v_p(e \cdot s) = u + v_p(s)$; for $2 \leq i \leq s$,

$$
\begin{aligned}
v_p(e^i \cdot C_s^i) &= iu + v_p(s(s-1)\cdots(s-i+1)) - v_p(i!) \\
&= iu + v_p(s) - v_p(i) + v_p(C_{s-1}^{i-1}) \\
&\geq iu + v_p(s) - v_p(i) \\
&= (u + v_p(s)) + (i-1)u - v_p(i) \\
&\geq (u + v_p(s)) + i - v_p(i) > u + v_p(s).
\end{aligned}
$$

Therefore, $v_p(d^s - 1) = u + v_p(s)$. The proof is completed.

Now return to the original problem. Let $a_1 = \varphi(n^2)$, where φ is Euler's function. According to the Euler theorem we know $n^2 \mid m^{a_1} - 1$. Let

$$d = m^{a_1}, \quad n = p_1^{\alpha_1} p_2^{\alpha_2} \cdots p_k^{\alpha_k}, \quad d - 1 = p_1^{\beta_1} p_2^{\beta_2} \cdots p_k^{\beta_k} N,$$

where $(n, N) = 1$. Then $\beta_i \geq 2\alpha_i \geq 2$ $(i = 1, 2, \ldots, k)$. Let b be a positive integer, satisfying $b\alpha_i \geq \beta_i$ $(i = 1, 2, \ldots, k)$, and let

$$a = a_1 p_1^{b\alpha_1 - \beta_1} p_2^{b\alpha_2 - \beta_2} \cdots p_k^{b\alpha_k - \beta_k}.$$

Then for $i = 1, 2, \ldots, k$, by the lemma, we have

$$v_{p_i}(m^a - 1) = v_{p_i}(d^{a/a_1} - 1) = \beta_i + (b\alpha_i - \beta_i) = b\alpha_i,$$

Therefore, $c = \dfrac{m^a - 1}{n^b}$ is a positive integer coprime with n. Here a, b, c meet the required condition. The proof is then completed. $\qquad\square$

4 Let n be a positive integer, $a_1, \ldots, a_n \in \{0, 1, \ldots, n\}$, and for $j(1 \le j \le n)$ define b_j as the number of elements contained in $\{i \mid i \in \{1, \ldots, n\}, a_i \ge j\}$. For example: when $n = 3$, if $a_1 = 1$, $a_2 = 2$, $a_3 = 1$, then $b_1 = 3$, $b_2 = 1$, $b_3 = 0$.

(1) Prove: $\sum_{i=1}^{n} (i + a_i)^2 \ge \sum_{i=1}^{n} (i + b_i)^2$.

(2) Prove: for $k \ge 3$, $\sum_{i=1}^{n} (i + a_i)^k \ge \sum_{i=1}^{n} (i + b_i)^k$.

Solution 1 Let $c_1 \ge \cdots \ge c_n$ be a rearrangement of $a_1, \ldots, a_n$ in descending order. We first prove:

$$\{1 + c_1, \ldots, n + c_n\} = \{1 + b_1, \ldots, n + b_n\}. \tag{1}$$

Construct an $n \times n$ grid table with four vertexes $(0, 0)$, $(0, n)$, (n, n), $(n, 0)$ in the plane rectangular coordinate system.

For each $i = 1, \ldots, n$, blacken the first c_i cells in the i-th column from bottom to top. Then by the definition of b_j we know that the first b_j cells in the j-th row from left to right are black, and the remaining are unpainted.

Starting from the upper left corner of the grid table $(0, n)$, draw a dividing line between the black grids and the unpainted grids until the lower right corner $(n, 0)$. Denote the drawn polyline by L, which consists of n horizontal line segments with length 1 and n vertical line segments with length 1. The sum of the horizontal and vertical coordinates of the midpoint of each horizontal line segment from left to right is $1 + c_1 - \frac{1}{2}, \ldots, n + c_n - \frac{1}{2}$, and the sum of the horizontal and vertical coordinates of the upper endpoint of each vertical line segment from bottom to top is $1 + b_1 - \frac{1}{2}, \ldots, n + b_n - \frac{1}{2}$.

For any positive integer r, draw straight line l_r: $x + y = r - \frac{1}{2}$. As $(0, n)$ and $(n, 0)$ must be on the same side of the line, the number of times the polyline L crosses l_r in the horizontal direction is equal to that in the vertical direction. Moreover, when the polyline and l_r intersect, they must intersect at the midpoint of a line segment of length 1. By the arbitrariness of r, we then get

$$\left\{1 + c_1 - \frac{1}{2}, \ldots, n + c_n - \frac{1}{2}\right\} = \left\{1 + b_1 - \frac{1}{2}, \ldots, n + b_n - \frac{1}{2}\right\}.$$

Therefore Equation (1) holds.

Since $c_1, \ldots, c_n$ and $1, \ldots, n$ are in reverse order, we have

$$\sum_{i=1}^{n}(i + a_i)^k = \sum_{i=1}^{n}\sum_{s=0}^{k} C_k^s i^s a_i^{k-s} = \sum_{s=0}^{k}\sum_{i=1}^{n} C_k^s i^s a_i^{k-s}$$

$$\geq \sum_{s=0}^{k}\sum_{i=1}^{n} C_k^s i^s c_i^{k-s} \ \text{(by sequence inequality)}$$

$$= \sum_{i=1}^{n}\sum_{s=0}^{k} C_k^s i^s c_i^{k-s} = \sum_{i=1}^{n}(i + c_i)^k = \sum_{i=1}^{n}(i + b_i)^k.$$

When $k = 2$, it is the proof of the first question; when $k \geq 3$, it is the proof of the second question.

Solution 2 (1) Notice that $b_j = \sum_{i, a_i \geq j} 1$. We have

$$\sum_{j=1}^{n}(j + b_j)^2 = \sum_{j=1}^{n} b_j^2 + 2\sum_{j=1}^{n} jb_j + \sum_{j=1}^{n} j^2$$

$$= \sum_{j=1}^{n}\sum_{a_{i_1} \geq j}\sum_{a_{i_2} \geq j} 1 + 2\sum_{j-1}^{n} j \sum_{a_i \geq j} 1 + \sum_{j=1}^{n} j^2$$

$$= \sum_{i_1, i_2}\sum_{j \geq \min(a_{i_1}, a_{i_2})} 1 + 2\sum_{i=1}^{n}\sum_{j \leq a_i} j + \sum_{j=1}^{n} j^2$$

$$= \sum_{i_1, i_2} \min(a_{i_1}, a_{i_2}) + 2\sum_{i=1}^{n}(1 + \cdots + a_i) + \sum_{j=1}^{n} j^2$$

$$= \sum_{i=1}^{n} a_i + 2\sum_{i_1 < i_2} \min(a_{i_1}, a_{i_2}) + \sum_{i=1}^{n} a_i(a_i + 1) + \sum_{j=1}^{n} j^2$$

$$\leq \sum_{i=1}^{n} a_i + 2\sum_{i_1 < i_2} a_{i_2} + \sum_{i=1}^{n} a_i(a_i + 1) + \sum_{j=1}^{n} j^2$$

$$= \sum_{i=1}^{n} a_i + 2\sum_{i_2=1}^{n}\sum_{i_1 < i_2} a_{i_2} + \sum_{i=1}^{n} a_i(a_i + 1) + \sum_{j=1}^{n} j^2$$

$$= \sum_{i=1}^{n} a_i + 2\sum_{i_2=1}^{n}(i_2 - 1)a_{i_2} + \sum_{i=1}^{n} a_i(a_i + 1) + \sum_{j=1}^{n} j^2$$

$$= \sum_{i=1}^{n}(i + a_i)^2.$$

(2) Rearrange $a_1, \ldots, a_n$ into $\bar{a}_1 \geq \cdots \geq \bar{a}_n$. By the definition of b_j we know

$$\bar{a}_{i \geq j} \text{ if and only if } i \leq b_j.$$

Therefore,

$$S = \{(i,j) \,|\, 1 \leq i \leq n, \ 1 \leq j \leq \bar{a}_i\} = \{(i,j) \,|\, 1 \leq j \leq n, \ 1 \leq i \leq b_j\}.$$

For any one variable function $f(x)$, let

$$F = \sum_{(i,j) \in S} (f(i+j) - f(i+j-1)).$$

We calculate the value of F with the two definition of set S, respectively.

For $S = \{(i,j) \,|\, 1 \leq i \leq n, \ 1 \leq j \leq \bar{a}_i\}$, we have

$$F = \sum_{i=1}^{n} \sum_{j=1}^{\bar{a}_i} (f(i+j) - f(i+j-1)) = \sum_{i=1}^{n} (f(i+\bar{a}_i) - f(i)).$$

For $S = \{(i,j) \,|\, 1 \leq j \leq n, \ 1 \leq i \leq b_j\}$, we have

$$F = \sum_{j=1}^{n} \sum_{i=1}^{b_j} (f(i+j) - f(i+j-1)) = \sum_{j=1}^{n} (f(j+b_j) - f(j)).$$

Therefore, we have

$$\sum_{i=1}^{n} f(i+\bar{a}_i) = \sum_{i=1}^{n} f(i+b_i).$$

Let $f(x) = x^k$. Then

$$\sum_{i=1}^{n} (i+\bar{a}_i)^k = \sum_{i=1}^{n} (i+b_i)^k.$$

Finally, using the sequence inequality for k, we have

$$\sum_{i=1}^{n} (i+b_i)^k = \sum_{i=1}^{n} (i+\bar{a}_i)^k = \sum_{i=1}^{n} \sum_{t=0}^{k} C_k^t i^{k-t} a_i^{-t} = \sum_{t=0}^{k} \sum_{i=1}^{n} C_k^t i^{k-t} a_i^{-t}$$

$$\leq \sum_{t=0}^{k} \sum_{i=1}^{n} C_k^t i^{k-t} a_i^t = \sum_{i=1}^{n} \sum_{t=0}^{k} C_k^t i^{k-t} a_i^t = \sum_{i=1}^{n} (i+a_i)^k.$$

The proofs are then completed. $\qquad\qquad\square$

Second Day
(8:00 – 12:00; August 13, 2016)

5 Let $S_n = a_1 + \cdots + a_n$ be the sum of the first n terms of sequence $a_1, a_2, \ldots$, and satisfy

$$S_1 = 1, \quad S_{n+1} = \frac{(2 + S_n)^2}{4 + S_n}, \quad n \geq 1.$$

Prove: for any positive integer n, $a_n \geq \dfrac{4}{\sqrt{9n + 7}}$.

Solution Since $S_n = S_n + \dfrac{4}{4 + S_n}$, we know $\{S_n\}$ is an increasing positive number sequence.

When $n = 1$, $a_1 = 1 = \dfrac{4}{\sqrt{9 \times 1 + 7}}$, the statement is true. So we just consider the case of $n \geq 2$ below.

Since $a_n = S_n - S_{n-1} = \dfrac{4}{4 + S_{n-1}}$, we just need to prove

$$4 + S_{n-1} \leq \sqrt{9n + 7}. \qquad\qquad ①$$

When $n = 2$, $4 + S_1 = 4 + 1 = 5 = \sqrt{9 \times 2 + 7}$, so ① holds.
Now assume $4 + S_{n-1} \leq \sqrt{9n + 7}$. We have

$$(4 + S_n)^2 = \left(4 + S_{n-1} + \frac{4}{4 + S_{n-1}}\right)^2$$

$$= (4 + S_{n-1})^2 + 8 + \frac{16}{(4 + S_{n-1})^2}$$

$$\leq 9n + 7 + 8 + 1 = 9(n + 1) + 7.$$

Therefore, $4 + S_n \leq \sqrt{9(n + 1) + 7}$. The proof by mathematical induction is then completed. $\qquad\square$

6 Find the largest positive integer m, so that each cell of an $m \times 8$ grid table can be filled with one of the four letters C, G, M, O, and has the following properties: there is at most one column in two different rows, so that the letters in the two rows are the same.

Solution The largest m is 5. First we give the following 5×8 table, satisfying the requirement in the problem.

C	C	C	C	G	G	G	G
C	G	G	G	C	C	C	M
G	C	M	M	C	M	M	C
M	M	C	O	M	C	O	C
O	O	O	C	O	O	C	O

Then we will prove, for any $m \geq 6$, the requirement cannot be met. As a fact of matter, if we can prove it in the case of $m = 6$, then the case of $m > 6$ is automatically proved, because at this time any $m \times 8$ table contains a 6×8 sub table enough to support the proof. Therefore in the following we consider the case $m = 6$.

For any column in a 6×8 table, let l be the number of pairs of rows that contain the same letter of C, G, M, O in this column. It is easy to see that

$$l \geq m - 4 = 2.$$

Then in the table there are at least $2 \times 8 = 16$ pairs of rows, such that each pair of rows among them contain the same letter in at least one column. Since $16 > C_6^2$ (the number of total row pairs in the table), then at least two such pairs of rows are the same, i.e. there is a pair of rows that contain the same letter in each of more than one column, defying the requirement.

Therefore, the largest positive integer is $m = 5$. $\qquad\square$

7 As shown in Fig. 7.1, in triangle ABC, $AB > AC$, I is its incenter, and D is the foot of I on the side BC. Through A draw $AH \perp BC$ at H, intersecting with BI and CI at P and Q, respectively. O is the circumcenter of triangle IPQ. Extend AO to intersect with BC at L. Let N be the second intersection point of line BC and the circumcircle of triangle AIL.

Prove: $\dfrac{BD}{CD} = \dfrac{BN}{CN}$.

Solution 1 As shown in Fig. 7.2, the projections of I on the sides of AC and AB are E and F, respectively, connecting DE, EF, and FD, and extending FE to intersect with the extension line of BC at N' and connecting IN'. It is easy to see that D, E, F are the tangent points of the inscribed circle

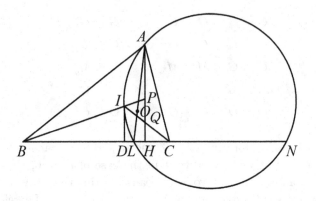

Fig. 7.1

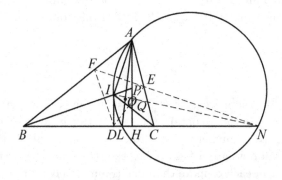

Fig. 7.2

of triangle ABC on the three sides. Then

$$ID = IE = IF, BF = BD,$$

$$CD = CE, AE = AF.$$

Note that

$$\angle DFE = \pi - \angle DFB - \angle EFA$$

$$= \pi - \frac{\pi - \angle ABC}{2} - \frac{\pi - \angle BAC}{2} = \frac{\pi - \angle ACB}{2}$$

$$= \frac{\pi}{2} - \angle ICB = \angle CQH = \angle IQP,$$

In the same way, $\angle DEF = \angle IPQ$. Therefore, $\triangle DEF \backsim \triangle IPQ$.

From
$$\angle AIP = \angle IAB + \angle IBA = \frac{\angle BAC + \angle ABC}{2} = \frac{\pi - \angle ACB}{2} = \angle IQP,$$
we know AI is the tangent to the circumscribed circle of $\triangle IPQ$ at I, with A being the intersection point of the tangent and line PQ.

On the other hand, from $ID \perp BC$ we know that BC is the tangent to the circumscribed circle of $\triangle DEF$ at point D, with N' being the intersection point of the tangent and line EF.

Moreover, O and I are the circumcenters of $\triangle IPQ$ and $\triangle DEF$, respectively, so $\angle IAO = \angle DN'I$. Therefore, A, I, L, N' are concyclic points. Because a straight line and a circle have only two intersection points, So N and N' coincide.

According to Menelaus' Theorem, $\dfrac{AF}{FB} \cdot \dfrac{BN}{NC} \cdot \dfrac{CE}{EA} = 1$. Therefore,
$$\frac{BN}{NC} = \frac{AE}{EC} \cdot \frac{BF}{FA} = \frac{BF}{EC} = \frac{BD}{DC}.$$

The proof is completed.

Solution 2 By the same way used in Solution 1 above, we get that AI is the tangent to the circumscribed circle of $\triangle IPQ$ at point I, so $\angle AIO = 90°$.

As shown in Fig. 7.3, let M be the midpoint of PQ. It is easy to know that $OM \perp PQ$, so $\angle AMO = 90°$, and then $A, I, O,$ and M are concyclic points; therefore, $\angle IMO = \angle IAO = \angle INL$.

OM and LN are parallel, as both they are perpendicular to AH. Combining it with $\angle IMO = \angle INL$, we know IM and IN are coincident, so $I, M,$ and N are collinear.

Since $ID \parallel PQ$, and the intersection point M of IN and PQ is exactly the midpoint of segment PQ, whose end points P and Q are also the intersection

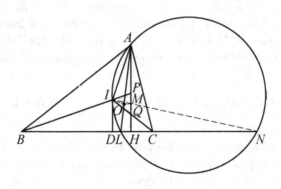

Fig. 7.3

points of IB, IC and PQ, respectively. So IB, ID, IC, and IN are harmonic pencil of lines, and then B, D, C, and N are harmonic range of points. Therefore, $\dfrac{BD}{DC} = \dfrac{BN}{NC}$.

The proof is completed. (The last paragraph can also be illustrated by a triangle) □

8 Let $\mathbb{Q}$ be the set of all rational numbers and $\mathbb{Z}$ be the set of all integers. On the coordinate plane, for a positive integer m, define

$$A_m = \left\{ (x, y) \mid x, y \in \mathbb{Q}, xy \neq 0, \frac{xy}{m} \in \mathbb{Z} \right\}.$$

For a line segment MN with M and N as the end points, define $f_m(MN)$ as the number of points on MN that belong to A_m.

Find the smallest real number λ, so that for any straight line l on the coordinate plane, a real number $\beta(l)$ exists, which satisfies: for any two points M and N on l,

$$f_{2016}(MN) \leq \lambda f_{2015}(MN) + \beta(l).$$

Solution We will prove that the smallest $\lambda = \dfrac{2015}{6}$. We first show that $\lambda = \dfrac{2015}{6}$ satisfies the given condition.

We call a point rational if its horizontal and vertical coordinates are both rational. We start with a few special cases.

If there is no more than one rational point on l, then just let $\beta(l) = 1$.

If l is a coordinate axis, then no point in any A_m on it, just let $\beta(l) = 0$.

If l is parallel to a coordinate axis, by symmetry we just need consider the case $l: x = r (r \in \mathbb{R}^+)$. When r is irrational, it is reduced to the previous case. When r is rational, for any pair of points $M(r, m)$, $N(r, n)$ on l, $m < n$, $f_{2016}(MN)$ is the number of multiples of 2016 in interval $[mr, nr]$, and $f_{2015}(MN)$ is that of 2015 in $[mr, nr]$. Since there exists at least one multiple of 2015 between any two consecutive multiples of 2016, then

$$f_{2016}(MN) \leq f_{2015}(MN) + 1 \leq \lambda f_{2015}(MN) + 1.$$

Therefore, just let $\beta(l) = 1$.

Now we assume that l is neither parallel nor coincident with any coordinate axis, and there are at least two rational points on it. Then it is easy to know that the equation of l can be written as $ax + by = c$, where $a, b, c \in \mathbb{Z}$, and $ab \neq 0$. By symmetry, we may further assume $a, b \in \mathbb{N}^+$.

For any positive integer m, if point (x, y) in A_m is on l, it should satisfy:

$$ax + by = c, \ x, y \in \mathbb{Q}, \quad \text{and} \quad \frac{xy}{m} \in \mathbb{Z}.$$

Then it is easy to find that $ax, by \in \mathbb{Z}$. If (x, y) is on l, then $(x + k \cdot 2015 \cdot 2016b, y - k \cdot 2015 \cdot 2016a)$ is also on l for any integer k, and furthermore,

$$(x, y) \in A_i \iff (x + k \cdot 2015 \cdot 2016b, y - k \cdot 2015 \cdot 2016a) \in A_i$$

($i = 2015, 2016$). So we let $\beta(l)$ be the number of points on l for $0 \leq x < 2015 \cdot 2016b$ that belong to A_{2016}. Define

$$g_{2015}(l) = |\{u \,|\, 0 \leq u < 2015 \cdot 2016ab, \ u \in \mathbb{Z}, \ 2015ab \,|\, u(c - u)\}|,$$

$$g_{2016}(l) = |\{u \,|\, 0 \leq u < 2015 \cdot 2016ab, \ u \in \mathbb{Z}, \ 2016ab \,|\, u(c - u)\}|.$$

($u = ax$). Then in an entire period, there are $g_i(l)$ points in A_i ($i = 2015, 2016$). We will prove that $g_{2016}(l) \leq \dfrac{2015}{6} g_{2015}(l)$ in the following. Let

$$2015 \cdot 2016ab = 2^{\alpha_1} \cdot 3^{\alpha_2} \cdot 7^{\alpha_3} \cdot 5^{\alpha_4} \cdot 13^{\alpha_5} \cdot 31^{\alpha_6} \cdot p_7^{\alpha_7} \cdot \ldots \cdot p_s^{\alpha_s}.$$

For any $p_i^k \,|\, 2015 \cdot 2016ab$, define $h(p_i^k)$ as the number of positive integers u in $0, 1, \ldots, p_i^{\alpha_i} - 1$ that satisfy $p_i^k \,|\, u(c - u)$. Then according to the Chinese remainder theorem,

$$g_{2016}(l) = h(2^{\alpha_1})h(3^{\alpha_2})h(7^{\alpha_3})h(5^{\alpha_4 - 1})h(13^{\alpha_5 - 1})h(31^{\alpha_6 - 1}) \prod_{i=7}^{s} h(p_i^{\alpha_i}),$$

$$g_{2015}(l) = h(2^{\alpha_1 - 5})h(3^{\alpha_2 - 2})h(7^{\alpha_3 - 1})h(5^{\alpha_4})h(13^{\alpha_5})h(31^{\alpha_6}) \prod_{i=7}^{s} h(p_i^{\alpha_i}).$$

Now we need prove

$$\frac{h(2^{\alpha_1})h(3^{\alpha_2})h(7^{\alpha_3})h(5^{\alpha_4 - 1})h(13^{\alpha_5 - 1})h(31^{\alpha_6 - 1})}{h(2^{\alpha_1 - 5})h(3^{\alpha_2 - 2})h(7^{\alpha_3 - 1})h(5^{\alpha_4})h(13^{\alpha_5})h(31^{\alpha_6})} \leq \frac{2015}{6}. \qquad \text{(1)}$$

For each $p_i^k \,|\, 2015 \cdot 2016ab$, assuming $p_i^d \,\|\, c$, if $k \leq 2d$, we have

$$p_i^k \,|\, u(c - u) \iff p_i^{[k/2]} \,|\, u,$$

then $h(p_i^k) = p_i^{\alpha_i - [k/2]}$; if $k > 2d$, we have

$$p_i^k \,|\, u(c - u) \iff p_i^{k-d} \,|\, u \text{ or } p_i^{k-d} \,|\, c - u \,|\, u,$$

then $h(p_i^k) = 2p_i^{\alpha_i - k + d}$. Therefore, the possible value of $\dfrac{h(p_i^k)}{h(p_i^{k+1})}$ is 1, $\dfrac{p}{2}$,

p; the possible value of $\dfrac{h(p_i^k)}{h(p_i^{k+2})}$ is $\dfrac{p}{2}$, p, $\dfrac{p^2}{2}$, p^2; and the possible value of

$\dfrac{h(p_i^k)}{h(p_i^{k+5})}$ is p^2, $\dfrac{p^3}{2}$, p^3, $\dfrac{p^4}{2}$, $\dfrac{p^5}{2}$, p^5.

So we get

$$\frac{h(2^{\alpha_1})h(3^{\alpha_2})h(7^{\alpha_3})h(5^{\alpha_4-1})h(13^{\alpha_5-1})h(31^{\alpha_6-1})}{h(2^{\alpha_1-5})h(3^{\alpha_2-2})h(7^{\alpha_3-1})h(5^{\alpha_4})h(13^{\alpha_5})h(31^{\alpha_6})}$$

$$\leq \frac{1}{4} \cdot \frac{2}{3} \cdot 1 \cdot 5 \cdot 13 \cdot 31 = \frac{2015}{6}.$$

The inequality ① is then proved. To prove that $\lambda = \dfrac{2015}{6}$ is the smallest,
let $ab = 21$, $c = 4 \times 3 \times 7 \times 5 \times 13 \times 31 = 84 \times 2015$, and $l : x + 21y = 84 \times 2015$.
We can check that the equality in ① holds.

The proof is completed.   □

China Girls' Mathematical Olympiad

2017 (Chongqing)

The 16th China Girls' Mathematical Olympiad (CGMO) was held between August 10th and 15th, 2017 at Chongqing No. 8 Secondary School, Chongqing, China. A total of 145 female students from 37 teams from various provinces, municipalities, and autonomous regions across the mainland China, Hong Kong, Macau, Russia, the Philippines, and Singapore participated in the competition. After two rounds of contests (4 hours for 4 questions each), 35 students headed by Feng Yuyang won the gold medal (first prize), 56 students headed by He Yuchen won the silver medal (second prize), and 54 students headed by Wu Junyan won the bronze medal (third prize). In addition, the top 12 contestants from the mainland China among those who won the gold medal are invited to participate in the National Mathematics Winter Camp (CMO) to be held in December 2017.

Professor Chen Min, vice president of the Academy of Mathematics and Systems Science, Chinese Academy of Sciences, vice chairman of the Chinese Mathematical Society and director of the Olympic Committee of the Chinese Mathematical Society, serves as the director of the organizing committee.

Director of the main examination committee: Li Shenghong.

Members of the main examination committee (in alphabetical order):

Ai Yinghua (Tsinghua University);

Bian Hongping (Huabo Electromechanical Co., Ltd.);

Chen Zi'ang (Peking University);

Fu Yunhao (Guangdong Second Normal University);

Ji Chungang (Nanjing Normal University);

Li Shenghong (Zhejiang University);

Liang Yingde (University of Macau);

Wang Bin (Chinese Academy of Sciences);

Wang Xinmao (University of Science and Technology of China).

First Day
(8:00 – 12:00; August 13, 2017)

1 (1) Find all positive integers n, such that for any positive odd integer $a, a^n - 1$ is divisible by 4.

(2) Find all positive integers n, such that for any positive odd integer a, $a^n - 1$ is divisible by 2^{2017}.

Solution (1) When n is even, $a^n \equiv 1 \pmod{4}$; when n is odd,

$$a^n - 1 \equiv (-1)^n - 1 \pmod{4}.$$

Thus the required n is necessarily and sufficiently a positive even number.

(2) When $2^{2015} \mid n$, let $n = 2^{2015} n_0$, and $b = a^{n_0}$. We have

$$a^n - 1 = b^{2^{2015}} - 1 = (b^2 - 1)(b^2 + 1)(b^4 + 1) \cdots (b^{2013} + 1)(b^{2014} + 1).$$

The first factor is a multiple of 8, and the other 2014 factors are all even. Hence

$$2^{2017} \mid a^n - 1,$$

and n satisfies the given condition.

Conversely, we show if 2^{2015} does not divide n, then $a^n - 1$ may not be divisible by 2^{2017}. Let

$$n = 2^t (2s + 1) (t, s \in Z \geq 0, \ t \leq 2014).$$

If $t = 0$, take $a = 3$,

$$a^n - 1 \equiv 3 - 1 \equiv 2 \pmod{4},$$

indicating n does not meet the requirement. If $1 \leq t \leq 2014$, again take $a = 3$,

$$a^n - 1 = 3^n - 1 = (3^{2^t} - 1)(3^{2^t \cdot 2s} + 3^{2^t \cdot (2s-1)} + \cdots + 3^{2^t} + 1)$$

$$= (3^{2^{t-1}} + 1)(3^{2^{t-2}} + 1) \cdots (3^2 + 1)(3^2 - 1)$$

$$(3^{2^t \cdot 2s} + 3^{2^t \cdot (2s-1)} + \cdots + 3^{2^t} + 1).$$

The first $t - 1$ factors are even but not divisible by 4; $3^2 - 1 = 2^3$; and the last factor is odd. It follows that $a^n - 1$ has 2 to the power $t + 2 \leq 2016$, and hence n is undesirable. □

2 As shown in Fig. 2.1 below, $ABCD$ is a convex quadrilateral such that

$$\angle BAD + 2\angle BCD = 180°.$$

The angle bisector of $\angle BAD$ intersects BD at E; the perpendicular bisector of AE intersects lines
CB, CD at points X, Y, respectively. Prove that A, X, C, Y are concyclic.

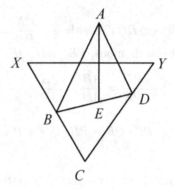

Fig. 2.1

Solution 1 As shown in Fig. 2.2, let XY intersect AB, AD at points P, Q, respectively. Connect EP, EQ. Since PQ is perpendicular to AE, the bisector of $\angle BAD$, it follows $\angle APQ = \angle AQP$, and $AP = AQ$. Since P, Q both lie on the perpendicular bisector of AE, $AP = EP$, $AQ = EQ$. Therefore, $APEQ$ has equal side lengths, and $APEQ$ is a rhombus. By

$$\angle PXB = 180° - \angle C - \angle QYC$$

$$= 90° + \angle EAD - \angle QYC$$

$$= \angle AQY - \angle QYC = \angle QDY,$$

and

$$\angle XPB = \angle APQ = \angle AQP = \angle DQY,$$

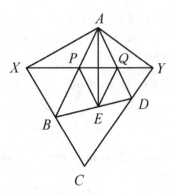

Fig. 2.2

we deduce that $\triangle XPB \backsim \triangle DQY$, and hence $\dfrac{PX}{PB} = \dfrac{QD}{QY}$, or $PX \cdot QY = PB \cdot QD$. Since $APEQ$ is a rhombus, $PE//AD$, $QE//AB$, and we have

$$\frac{PB}{PE} = \frac{AB}{AD} = \frac{QE}{QD},$$

which implies $PA \cdot QA = PE \cdot QE = PB \cdot QD = PX \cdot QY$, or $\dfrac{PA}{PX} = \dfrac{QY}{QA}$. Moreover,

$$\angle APX = 180° - \angle APQ = 180° - \angle AQP = \angle YQA,$$

and thus $\triangle APX \backsim \triangle YQA$. We obtain

$$\angle XAY = \angle XAQ + \angle YAQ$$
$$= \angle XAQ + \angle AXQ$$
$$= 180° - \angle AQX,$$

and $\angle C = 90° - \angle EAD = \angle AQX$. Therefore $\angle XAY + \angle C = 180°$, and A, X, C, Y are concyclic.

Solution 2   As shown in Fig. 2.3, draw altitudes from point A to lines XY, BC, CD: L, M, N being the feet, respectively. Let U, V be the symmetric points of A about lines BC, CD, respectively. To show that A, X, C, Y are concyclic, by the inverse Simson theorem, it suffices to show L, M, N are collinear. Furthermore, since L, M, N are midpoints of AE, AU, AV, respectively, it suffices to show E, U, V are collinear.

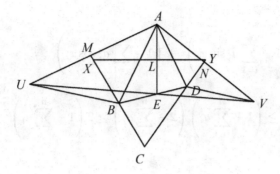

Fig. 2.3

From
$$\frac{BU}{BE} = \frac{AB}{BE} = \frac{AD}{DE} = \frac{DV}{DE},$$
and
$$\angle UBX + \angle XBD + \angle BDY + \angle YDV = \angle ABX + 180° + \angle BCD + \angle ADY$$
$$= 540° + \angle BCD - \angle ABC - \angle ADC$$
$$= 180° + \angle BAD + 2\angle BCD$$
$$= 360°,$$

we deduce that U, V are on opposite sides of BD, and $\angle UBE = \angle VDE$. Therefore $\triangle UBE \backsim \triangle VDE$, $\angle BEU = \angle DEV$, and it follows that E, U, V are collinear, completing the proof. $\square$

3 Let $a_i \geq 0$, $x_i \in \mathbb{R}$, $i = 1, 2, \ldots, n$. Prove,
$$\left[\left(1 - \sum_{i=1}^{n} a_i \cos x_i\right)^2 + \left(1 - \sum_{i=1}^{n} a_i \sin x_i\right)^2\right]^2 \geq 4\left(1 - \sum_{i=1}^{n} a_i\right)^3.$$

Solution 1 If $1 - \sum\limits_{i=1}^{n} a_n \leq 0$, the statement is obviously true. If $1 - \sum\limits_{i=1}^{n} a_n \geq 0$, we have
$$\left(1 - \sum_{i=1}^{n} a_i \sin x_i\right)^2 \geq \left(1 - \sum_{i=1}^{n} a_i\right)\left(1 - \sum_{i=1}^{n} a_i \sin^2 x_i\right),$$
$$\left(1 - \sum_{i=1}^{n} a_i \cos x_i\right)^2 \geq \left(1 - \sum_{i=1}^{n} a_i\right)\left(1 - \sum_{i=1}^{n} a_i \cos^2 x_i\right).$$

Then

$$\left(1 - \sum_{i=1}^{n} a_i \sin x_i\right)^2 + \left(1 - \sum_{i=1}^{n} a_i \cos x_i\right)^2$$

$$\geq \left(1 - \sum_{i=1}^{n} a_i\right)\left(2 - \sum_{i=1}^{n} a_i\right) \geq 2\left(1 - \sum_{i=1}^{n} a_i\right)^{\frac{3}{2}}.$$

It follows

$$\left[\left(1 - \sum_{i=1}^{n} a_i \cos x_i\right)^2 + \left(1 - \sum_{i=1}^{n} a_i \sin x_i\right)^2\right]^2 \geq 4\left(1 - \sum_{i=1}^{n} a_i\right)^3.$$

Solution 2 Let $\alpha = (1, 1)$, $\alpha_k = (-a_k \sin x_k, -a_k \cos x_k)$, $k = 1, 2, \ldots, n$. We have

$$\left|\alpha + \sum_{k=1}^{n} \alpha_k\right| \geq |\alpha| - \sum_{k=1}^{n} |\alpha_k| = \sqrt{2} - \sum_{k=1}^{n} a_k.$$

If $\sum_{k=1}^{n} a_k > 1$, the statement is obviously true. If $\sum_{k=1}^{n} a_k = r \leq 1$,

$$\left(1 - \sum_{i=1}^{n} a_i \sin x_i\right)^2 + \left(1 - \sum_{i=1}^{n} a_i \cos x_i\right)^2 \geq (\sqrt{2} - r)^4.$$

Now it suffices to prove for $0 \leq r \leq 1$, $(\sqrt{2} - r)^4 \geq 4(1 - r)^3$. This is equivalent to

$$r^4 - 4\sqrt{2}r^3 + 12r^2 - 8\sqrt{2}r + 4$$

$$\geq -4r^3 + 12r^2 - 12r + 4$$

$$\Leftrightarrow f(r) = r^3 - 4(\sqrt{2} - 1)r^2 + 12 - 8\sqrt{2} \geq 0.$$

Since $f'(r) = 3r^2 - 8(\sqrt{2} - 1)r \leq 0$, we have

$$f(r) \geq f(1) = (\sqrt{2} - 1)^4 > 0. \qquad \square$$

④ Consider all pairs (S, T) of sets satisfying $S \cup T = \left\{\dfrac{1}{2002}, \dfrac{1}{2003}, \ldots, \dfrac{1}{2017}\right\}$, and $|S| = |T| = 8$. Let A, B be the sums of all elements of S and T, respectively. Determine all possible pair(s) (S, T) such that the value of $|A - B|$ is minimal.

Solution (The following proof is rewritten by Chen Haoran based on the original proof in the Chinese version of this book) Let $M = 2009.5$. Rephrase the problem as to partition $E = \{\pm 0.5, \pm 1.5, \ldots, \pm 7.5\}$ into two disjoint sets $A, B, |A| = |B| = 8$, such that

$$|S_A - S_B| := \left| \sum_{e \in A} \frac{1}{M+e} - \sum_{e \in B} \frac{1}{M+e} \right| \qquad (*)$$

is minimal. For $m = 1, 2, 3, 4$ and $E = A \cup B, |A| = |B| = 8$, define

$$A_m = \sum_{e \in A} e^m, \quad B_m = \sum_{e \in B} e^m, \quad E_m = \sum_{e \in E} e^m.$$

We have $E_1 = E_3 = 0$, $E_2 = 340$, $E_4 = 12937$ and $A_m + B_m = E_m$, $1 \le m \le 4$. For every $e \in E$, we shall use the following decompositions:

$$\frac{1}{M+e} = \frac{1}{M} \left(\frac{1}{1 + \dfrac{e}{M}} \right) = \frac{1}{M} \left(1 - \frac{e}{M} + \frac{e^2}{M(M+e)} \right) \qquad \text{①}$$

$$= \frac{1}{M} \left(1 - \frac{e}{M} + \frac{e^2}{M^2} - \frac{e^3}{M^2(M+e)} \right) \qquad \text{②}$$

$$= \frac{1}{M} \left(1 - \frac{e}{M} + \frac{e^2}{M^2} - \frac{e^3}{M^3} + \frac{e^4}{M^3(M+e)} \right). \qquad \text{③}$$

Define $R(T, m) = \sum_{e \in T} \dfrac{e^m}{M+e}$, and $R(A - B, m) = R(A, m) - R(B, m)$. Let $A = A^*$, $B = B^*$ minimizes $(*)$. We prove a few claims.

Claim 1 $A_1^* = B_1^*$.

Proof If not, then $|A_1^* - B_1^*| \ge 2$, because $A_1^* + B_1^* = E_1 = 0$ and they are integers. For any A, B satisfying $A_1 = B_1$, use ① to get

$$M|S_{A^*} - S_{B^*}| \ge 2 - |R(A^* - B^*, 2)| > 2 - \frac{E_2}{M - 7.5} > 1$$

$$> \frac{E_2}{M - 7.5} > R(A - B, 2) = M|S_A - S_B|.$$

This implies that A^*, B^* cannot minimize $(*)$.

Claim 2 $A_2^* = B_2^*$.

Proof   Define $G_3(t) = (t+8)(t-4)^2$, $0 < G_3(t) < 254$ for $t \in E$. For any A, B satisfying $|A| = |B|$ and $A_1 = B_1$, it is simple to check that

$$x := A_3 - B_3 = \sum_{t \in A} G_3(t) - \sum_{t \in B} G_3(t) < 2000. \qquad (1)$$

Now we can estimate

$$|R(A - B, 3)| = \left| \sum_{e \in A} \frac{e^3}{M+e} - \sum_{e \in B} \frac{e^3}{M+e} \right|$$

$$< \left| \sum_{e \in A, e > 0} \frac{e^3}{M-7.5} - \sum_{e \in B, e < 0} \frac{e^3}{M-7.5} + \sum_{e \in A, e < 0} \frac{e^3}{M+7.5} \right.$$

$$\left. - \sum_{e \in B, e > 0} \frac{e^3}{M+7.5} \right|$$

$$< \left| \frac{A_3 - B_3}{M-7.5} + \frac{15B_3}{(M-7.5)(M+7.5)} \right|$$

$$= \left| \frac{x}{M-7.5} + \frac{15(1016 - \frac{1}{2}x)}{(M-7.5)(M+7.5)} \right|$$

$$:= |f(x)| \cdot \left(\text{Here} \sum_{0.5}^{7.5} e^3 = 1016. \right)$$

Clearly, $f(x)$ is increasing, and it follows from (1) that $|f(x)| < |f(2000)| < 1$. If the claim is not true, then $|A_2^* - B_2^*| \geq 2$, and $x^* = A_3^* - B_3^*$ also satisfies (1), $|f(x^*)| < 1$. Assume $A_i = B_i (i = 0, 1, 2)$ for some A, B, and $|A^*| = |B^*|$, $A_1^* = B_1^*$, $|A_2^* - B_2^*| \geq 2$. By ②, we have

$$M^2 |S_{A^*} - S_{B^*}| > 2 - |R(A^* - B^*, 3)| > 1 > |R(A - B, 3)|$$

$$= M^2 |S_A - S_B|$$

and hence A^*, B^* cannot minimize (*). □

Claim 3 $A_3^* = B_3^*$.

Proof   Similar to the proof of Claim 2, define $G_4(t) = (t - 7.5)(t - 0.5)(t + 0.5)(t + 7.5)$, $-780 \leq G_4(t) \leq 0$. For any A, B satisfying $|A| = |B|$, $A_1 = B_1$

and $A_2 = B_2$,

$$y := A_4 - B_4 = \sum_{t \in A} G_4(t) - \sum_{t \in B} G_4(t) < 780 \cdot 8 < 6240. \qquad (2)$$

Estimating

$$|R(A - B, 4)| = \left| \sum_{e \in A} \frac{e^4}{M + e} - \sum_{e \in B} \frac{e^4}{M + e} \right|$$

$$< \left| \frac{A_4 - B_4}{M - 7.5} + \frac{15 B_4}{(M - 7.5)(M + 7.5)} \right|$$

$$= \left| \frac{y}{M - 7.5} + \frac{15(E_4 - y)}{2(M - 7.5)(M + 7.5)} \right| := |g(y)|,$$

from (2) we have $R(A - B, 4) < |g(6240)| < 4$, which also holds for $R(A^* - B^*, 4)$. Now, if $A_3^* \neq B_3^*$, observe $6 \mid t(t-1)(t+1)$ for $t \in E$. It follows that $\dfrac{A_3^* - A_1^*}{6}$ and $\dfrac{B_3^* - B_1^*}{6}$ are both integers, and $|A_3^* - B_3^*| \geq 12$ since their sum is even. Assume $A_i = B_i$ ($i = 0, 1, 2, \ldots, 3$) for $A, B, A_j^* = B_j^*$ ($j = 0, 1, 2$) and $|A_3^* - B_3^*| \geq 12$. By ③, we obtain

$$M^3 |S_{A^*} - S_{B^*}| > 12 - |R(A^* - B^*, 4)| > 8 > |R(A - B, 4)| = M^3 |S_A - S_B|,$$

and hence A^*, B^* cannot minimize (*).

Claim 4 *Let A' consist of all integers in $[1, 16]$ whose binary representations have odd number of 1's, and let B' consist of those with even number of 1's. Then A' and B' is the unique partition that satisfies $A_i' = B_i'$ for $i = 0, 1, 2, 3$.*

Proof In general, one can divide $[1, 2^n]$ into A' and B' in a similar manner and use induction to show $A_i' = B_i'$ for $i = 0, \ldots, n-1$. Here, we only prove the uniqueness.

Notice that $t^3 \equiv 0, 1, -1 \pmod 9$ when $t \equiv 0, 1, 2 \pmod 3$, and

$$\sum_1^{16} t^3 \equiv 1, \sum_1^{16} (t-1)^3 \equiv 0 \pmod 9$$

which implies

$$\sum_{t \in A} t^3 \equiv \sum_{t \in B} t^3 \equiv 5, \sum_{t \in A} (t-1)^3 \equiv \sum_{t \in B} (t-1)^3 \equiv 0 \pmod 9.$$

If we let a_i, b_i $(i = 0, 1, 2)$ be the number of elements in A' and B' that are congruent to 0, 1, 2 modulo 3, respectively, then

$$a_1 - a_2 \equiv 5, -a_0 + a_2 \equiv 0, b_1 - b_2 \equiv 5, -b_0 + b_2 \equiv 0 \pmod 9.$$

It follows that $a_0 = a_2, b_0 = b_2$ (since they are all ≤ 6), and by letting $a_1 = a_2 + 5$ we find $(a_0, a_1, a_2, b_0, b_1, b_2) = (1, 6, 1, 4, 0, 4)$. So A' includes $1, 4, 7, 10, 13, 16$, and from A'_1 and A'_2 it is easy to figure out that A' also includes 6 and 11. The uniqueness is now verified (by letting $a_1 = a_2 - 4$ we get the symmetric solution).

Finally, we let the two sets be $\left\{ \dfrac{1}{e + 2001} : e \in A' \right\}$ and $\left\{ \dfrac{1}{e + 2001} : e \in B' \right\}$. They minimize the difference desired in the original problem.

Remark 1   The idea of the decompositions ① ② ③ comes from Taylor expansion.

Remark 2   The functions $G_3(t), G_4(t)$ are carefully chosen such that the range of their values on E is minimal and the coefficients of the second highest order terms are 0, so we have estimates (1) and (2).

Remark 3   The Prouhet-Tarry-Escott problem asks whether it is possible to partition $[0, N]$ into two disjoint sets S_1 and S_2, such that

$$\sum_{t \in S_1} t^i = \sum_{t \in S_2} t^i \quad \text{for } i = 0, 1, \dots, k.$$

A sufficient but not necessary condition is that N is a multiple of 2^{k+1}, and Claim 4 is a special case when $N = 2^{k+1}$. The Prouhet-Tarry-Escott problem has close relation with the Thue-Morse sequence: it starts with 0 and successively appends the Boolean complement of the sequence obtained thus far, so the strings yielded are $0, 01, 0110, 01101001, 0110100110010110$, and so on. The n-th number indicates the parity of the number of 1's in the binary representation of n. It is worth noting that for general n, the proof of uniqueness of partition (up to symmetry) is not elementary.

Second Day
(8:00 – 12:00; August 14, 2017)

⑤   Determine the largest real number C, such that for any positive integer n and any sequence $\{x_k\}$ satisfying $0 = x_0 < x_1 < x_2 < \cdots < x_n = 1$, the following inequality always holds:

$$\sum_{k=1}^{n} x_k^2 (x_k - x_{k-1}) > C.$$

Solution First, we claim $C \geq \dfrac{1}{3}$. By $x_k \geq x_{k-1} \geq 0$. We have

$$3\sum_{k=1}^{n} x_k^2(x_k - x_{k-1}) = \sum_{k=1}^{n}(x_k^2 + x_k^2 + x_k^2)(x_k - x_{k-1})$$

$$> \sum_{k=1}^{n}(x_k^2 + x_k x_{k-1} + x_{k-1}^2)(x_k - x_{k-1})$$

$$= \sum_{k=1}^{n}(x_k^3 - x_{k-1}^3) = x_n^3 - x_0^3 = 1 - 0 = 1.$$

On the other hand, if $C > \dfrac{1}{3}$, there exist $0 = x_0 < x_1 < x_2 < \cdots < x_n = 1$ such that

$$\sum_{k=1}^{n} x_k^2(x_k - x_{k-1}) < C.$$

Indeed, let $x_k = \dfrac{k}{n}$, $k = 0, 1, \ldots, n$. For $n > 3$ and $\dfrac{1}{n} < C - \dfrac{1}{3}$, we have

$$\sum_{k=1}^{n} x_k^2(x_k - x_{k-1}) = \frac{1}{n^3}\sum_{k=1}^{n} k^2 = \frac{n(n+1)(2n+1)}{6n^3}$$

$$= \frac{1}{3} + \frac{1}{2n}\left(1 + \frac{3}{n}\right)$$

$$< \frac{1}{3} + \frac{1}{n} < C.$$

Hence $C = \dfrac{1}{3}$. $\qquad\qquad\qquad\qquad\qquad\qquad\qquad\qquad\qquad\qquad\square$

6 Let n be a positive integer, and X be a finite set. Let mapping $f\colon X \to X$ satisfy that for every $x \in X$, $f^{(n)}(x) = x$, in which $f^{(1)}(x) = f(x)$, $f^{(i)}(x) = f(f^{(i-1)}(x))$, $i \geq 2$. Denote m_j as the number of elements in $\{x \in X \mid f^{(j)}(x) = x\}$. If k is an integer, prove:

(a) $\dfrac{1}{n}\sum_{j=1}^{n} m_j \sin\dfrac{2jk\pi}{n} = 0$;

(b) $\dfrac{1}{n}\sum_{j=1}^{n} m_j \cos\dfrac{2jk\pi}{n}$ is a nonnegative integer.

Solution Let G be a directed graph in which all vertices are elements of X, and for every $x, y \in X$, there is a directed edge from x to y if and only if $f(x) = y$. If x is a fixed point of f, then there is an edge from x to itself, which is called an orbit of length 1. Since $f^{(n)}(x) = x$ for every $x \in X$, f is a bijection. Removing all fixed points of f and their edges, we obtain a graph in which every vertex has indegree and outdegree both equal to 1. It follows that the graph consists of disjoint cycles or orbits, of length l if it contains l vertices. Clearly, $f^{(j)}(x) = x$ if and only if l divides j; in particular, $f^{(n)}(x) = x$, and l divides n.

Suppose G has p orbits (including the orbits of length 1), whose lengths are $l_1, \ldots, l_p$, respectively. For each j, we have

$$m_j = \sum_{1 \le t \le p, l_t \mid j} l_t.$$

It follows, with interchanging the order of summations, that

$$\frac{1}{n} \sum_{j=1}^{n} m_j e^{\frac{2jk\pi i}{n}} = \frac{1}{n} \sum_{j=1}^{n} e^{\frac{2jk\pi i}{n}} \sum_{l_t \mid j} l_t = \frac{1}{n} \sum_{t=1}^{p} l_t \sum_{1 \le j \le n, l_t \mid j} e^{\frac{2jk\pi i}{n}}$$

$$= \frac{1}{n} \sum_{t=1}^{p} l_t \sum_{q=1}^{n/l_t} e^{\frac{2kq\pi i}{n/l_t}} = \frac{1}{n} \sum_{\frac{n}{(n,k)} \mid l_t} l_t \cdot \frac{n}{l_t} = \sum_{\frac{n}{(n,k)} \mid l_t} 1, \quad \text{(1)}$$

in which the last but one equality comes from the well known result

$$\sum_{q=1}^{n/l_t} e^{\frac{2kq\pi i}{n/l_t}} = \begin{cases} \dfrac{n}{l_t}, & \text{if } \dfrac{n}{(n,k)} \bigg| l_t, \\ 0, & \text{otherwise} \end{cases}.$$

Comparing the real and the imaginary parts on both sides of (1), we deduce (a) and (b).   □

7   As shown in Fig. 7.1, let ω_1 be the circumcircle of quadrilateral $ABCD$, AC intersect BD at E, and AD intersect BC at F. Let circle ω_2 touch segments EB, EC at M, N, respectively, and intersect ω_1 at Q, R. Let lines BC, AD intersect MN at S, T, respectively. Prove that Q, R, S, T are concyclic.

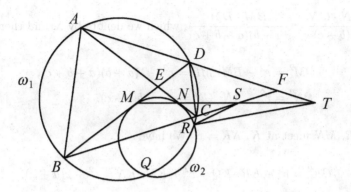

Fig. 7.1

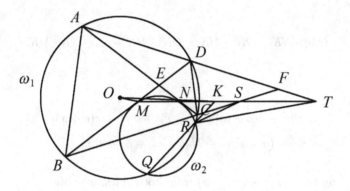

Fig. 7.2

Solution 1 As shown in Fig. 7.2, let $MN = a$, $NS = b$, $ST = c$, and let ω_1 be centered at O, with radius r. Since

$$\angle DTM = \angle ENM - \angle NAT$$
$$= \angle EMN - \angle MBS$$
$$= \angle CSN,$$

and $\angle DMT = \angle ENM = \angle CNT$, we have $\triangle DMT \backsim CNS$. Since

$$\angle ATN = \angle DTM = \angle CSN = \angle BSM, \angle TAN = \angle SBM,$$

we have $\triangle ANT \backsim \triangle BMS$. It follows

$$\frac{AN}{b+c} = \frac{BM}{a+b}, \frac{CN}{b} = \frac{DM}{a+b+c},$$

and $\dfrac{AN \cdot CN}{b(b+c)} = \dfrac{BM \cdot DM}{(a+b)(a+b+c)}$, which we denote by λ, and thereby

$$OM^2 = r^2 - BM \cdot MD = r^2 - \lambda(a+b)(a+b+c),$$
$$ON^2 = r^2 - AN \cdot CN = r^2 - \lambda b(b+c).$$

Let QR, MN meet at K, $NK = x$. We have

$$OK^2 = r^2 + KR \cdot KQ = r^2 + KM \cdot KN = r^2 + x(a+x).$$

Since N lies on MK of $\triangle OMK$ (possibly degenerate), by Stewart's theorem, we have

$$OM^2 \cdot NK + OK^2 \cdot MN = ON^2 \cdot MK + MN \cdot NK \cdot MK,$$

or

$$x[r^2 - \lambda(a+b)(a+b+c)] + a[r^2 + x(a+x)]$$
$$= (a+x)[r^2 - \lambda b(b+c)] + ax(a+x).$$

Subtracting $(a+x)r^2 + ax(a+x)$ from both sides, we obtain

$$-\lambda x(a+b)(a+b+c) = -\lambda(a+x)b(b+c).$$

Thus $x(a+b)(a+b+c) = (a+x)b(b+c)$, and

$$x = \frac{ab(b+c)}{(a+b)(a+b+c) - b(b+c)} = \frac{b(b+c)}{a+2b+c},$$

$$MK \cdot NK - SK \cdot TK = x(a+x) - (b-x)(b+c-x)$$
$$= x(a+2b+c) - b(b+c) = 0.$$

It follows $SK \cdot TK = MK \cdot NK = QK \cdot RK$, and Q, R, S, T are concyclic.

Solution 2 Notice that

$$\angle DTM = \angle ENM - \angle NAT = \angle EMN - \angle MBS = \angle CSN, \text{ and}$$

$$\angle FST = \angle CSN = \angle DTM = \angle FTS.$$

Hence $FS = FT$. By the law of sines,

$$\frac{AT}{AN} = \frac{\sin\angle ANT}{\sin\angle ATN}, \quad \frac{BS}{BM} = \frac{\sin\angle BMS}{\sin\angle BSM},$$

$$\frac{CS}{CN} = \frac{\sin\angle CNS}{\sin\angle CSN}, \quad \frac{DT}{DM} = \frac{\sin\angle DMT}{\sin\angle DTM}. \qquad \text{①}$$

From $\angle ATN = \angle DTM = \angle CSN = \angle BSM$, and

$$180° - \angle BMS = \angle DMT = \angle ANM = \angle CNS = 180° - \angle ANT,$$

we deduce that in ①, every fraction on the right hand side is equal, and thus

$$\frac{AT}{AN} = \frac{BS}{BM} = \frac{CS}{CN} = \frac{DT}{DM} := k.$$

Since $FS = FT$, there exists circle ω_3 that touches FS at S and FT at T. Now as shown in Fig. 7.3, let circles $\omega_1, \omega_2, \omega_3$ be centered at O_1, O_2, O_3 and be with radii r_1, r_2, r_3, respectively.

For point P on the plane, define function

$$f(P) = (k^2 - 1)(PO_1^2 - r_1^2) - k^2(PO_2^2 - r_2^2) + (PO_3^2 - r_3^2).$$

Then

$$f(A) = (k^2 - 1) \cdot 0 - k^2 \cdot AN^2 + AT^2 = 0,$$

and similarly $f(B) = f(C) = f(D) = 0$. In the Cartesian plane, let the coordinates of O_i be (x_i, y_i) $(i = 1, 2, 3)$. We have

$$f(x, y) = (k^2 - 1)[(x - x_1)^2 + (y - y_1)^2 - r_1^2]$$

$$- k^2[(x - x_2)^2 + (y - y_2)^2 - r_2^2]$$

$$+ [(x - x_3)^2 + (y - y_3)^2 - r_3^2]$$

$$= ax + by + c. \ (a, b, c \text{ are constants})$$

If $(a, b, c) \neq (0,0,0)$, $f(x, y) = 0$ represents a line (if not degenerate); however, it passes through A, B, C and D, a contradiction. This implies $a = b = c = d = 0$, and f is constantly equal to zero. By $RO_1^2 = r_1^2$, $RO_2^2 = r_2^2$ and $f(R) = 0$, we deduce $RO_3^2 = r_3^2$, and R is on ω_3. Likewise, Q is on ω_3. Therefore, Q, R, S, T are all on ω_3. $\qquad \square$

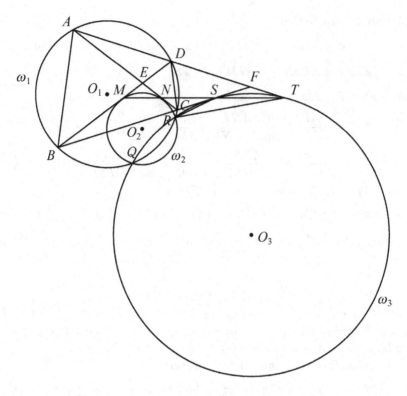

Fig. 7.3

8   Given a positive integer $n \geq 2$ and two $n \times n$ tables

$$A = \begin{bmatrix} a_{11} & a_{12} & \cdots & a_{1n} \\ a_{21} & a_{22} & \cdots & a_{2n} \\ \vdots & \vdots & \cdots & \vdots \\ a_{n1} & a_{n2} & \cdots & a_{nn} \end{bmatrix}, \quad B = \begin{bmatrix} b_{11} & b_{12} & \cdots & b_{1n} \\ b_{21} & b_{22} & \cdots & b_{2n} \\ \vdots & \vdots & \cdots & \vdots \\ b_{n1} & b_{n2} & \cdots & b_{nn} \end{bmatrix}$$

that satisfy $\{a_{ij} \mid 1 \leq i, j \leq n\} = \{b_{ij} \mid 1 \leq i,\ j \leq n\} = \{1, 2, \ldots, n^2\}$. Define a swap on A as interchanging two entries in the same row or in the same column of A and leaving the other $n^2 - 2$ entries unchanged. Find the minimum number m, such that given any tables A and B, one can change A into B by performing at most m swaps.

Solution   Without loss of generality, for $1 \leq i, j \leq n$, let $b_{ij} = (i-1)n + j$. We need the following two lemmas.

Lemma 1 *Any permutation* $(a_1, a_2, \ldots, a_n)$ *of* 1 *through* n *can be changed into* $(1, 2, \ldots, n)$ *by at most* $n - 1$ *swaps.*

For $i = 1, 2, \ldots, n - 1$, swap i and the i-th entry.

Lemma 2 *To change* $(2, 3, \ldots, n, 1)$ *into* $(1, 2, 3, \ldots, n)$, *at least* $n - 1$ *swaps are needed.*

Proof We use induction on n. Clearly the lemma is true when $n = 2$. Suppose that $m < n - 1$ swaps may change $(2, 3, \ldots, n, 1)$ into $(1, 2, 3, \ldots, n)$. Then for some $k \in \{2, 3, \ldots, n\}$, only one swap involves k, which moves k from position $k - 1$ to position k. However, this implies that $m - 1$ swaps may change $(2, \ldots, k - 1, k + 1, \ldots, n, 1)$ into $(1, 2, \ldots, k - 1, k + 1, \ldots, n)$, a contradiction. The lemma is verified.

Now we perform swaps on the $n \times n$ table A. First, change row 1 into $(1, 2, \ldots, n)$: there are two cases.

(a) If $1, 2, \ldots, n$ are in distinct columns, then at most n swaps are needed to move them into row 1, and another $n - 1$ swaps are needed to make them in the ascending order.

(b) If $1, 2, \ldots, n$ are not in distinct columns, then there exists a column, say column k, without any of these numbers. Use two swaps to move the number k into this column and then into row 1, and keep the positions of $1, \ldots, k - 1, k\_1, \ldots, n$ unchanged.

In either case, at most $2n - 1$ swaps are needed.

Second, change row i, $i = 2, 3, \ldots, n - 1$, into $((i - 1)n + 1, (i - 1)n + 2, \ldots, in)$. Every row requires at most $2n - 1$ swaps.

Finally, at most $n - 1$ swaps are needed to change the last row into $((n - 1)n + 1, (n - 1)n + 2, \ldots, n^2)$. The total number of swaps is $m = 2n(n - 1)$.

We must prove $2n(n - 1)$ is minimal. Let

$$
A = \begin{bmatrix}
n + 2 & n + 3 & \cdots & 2n & n + 1 \\
2n + 2 & 2n + 3 & \cdots & 3n & 2n + 1 \\
\vdots & \vdots & & \vdots & \vdots \\
(n - 1)n + 2 & (n - 1)n + 3 & \cdots & n^2 & (n - 1)n + 1 \\
2 & 3 & \cdots & n & 1
\end{bmatrix}.
$$

Notice that all entries in the i-th column of A are in the $(i+1)$-st column of B. By lemma 2, at least $n(n-1)$ swaps (in rows) are needed to have all the entries positioned into the correct columns; analogously, at least $n(n-1)$ swaps (in columns) are needed to have all the entries positioned into the correct rows. The whole process requires $2n(n-1)$ swaps.   □

China Western Mathematical Olympiad

2016 (Mianyang, Sichuan)

The 2016 China West Mathematical Olympiad (CWMO) was held in Mianyang, Sichuan from August 12th to 17th, hosted by Mianyang Middle School, sponsored by the Organizing Committee of the West China Mathematical Olympiad and Sichuan Mathematics Society.

A total of 231 students (81 formal and 150 non-formal contestants) from 21 teams, including the Hong Kong team, participated in this tournament. In addition, Kazakhstan, Singapore, Indonesia, the Philippines, and other countries also sent teams to participate in. The game lasted for two days, with 4 questions a day, each with 15 points, and a perfect score of 120 points. Amanbayeva Aruzhan from Kazakhstan tied with the other 5 students to get the personal first (gaining full points), and the Sichuan team won the first place in total score. In this competition, 28 students won the gold medals, 31 won the silver medals, and 22 won the bronze medals. Cheng Puhua from Singapore received the highest scores among non-formal contestants.

The director of the main test committee of this competition is Leng Gangsong (Professor of Shanghai University).

The members are:

Liang Yingde (Professor of University of Macau);

Xiong Bin (Professor of East China Normal University);

Zou Jin (Gao Si Education);

Zhang Xinze (Wugang No. 3 Middle School);

Yang Mingliang (Affiliated Middle School of Hunan Normal University);

Zhang Duanyang (Affiliated Middle School of People's University);
Wang Guangting (Shanghai Middle School);
Shi Zehui (Affiliated Middle School of Jilin University);
Shi Kejie (Affiliated Middle School of Fudan University);
Lin Tianqi (graduate student of East China Normal University);
Liu Shixiong (Zhongshan Campus, Middle School Attached to South China Normal University);
Feng Zhigang (Shanghai Middle School).

First Day
(8:00 – 12:00; August 15, 2016)

1 Real numbers a, b, c, d satisfy $abcd > 0$. Prove there is a permutation x, y, z, w of a, b, c, d such that $2(xz + yw)^2 > (x^2 + y^2)(z^2 + w^2)$. (Contributed by Liu Shixiong)

Solution 1 By the method of reduction to absurdity, we assume that for any permutation x, y, z, w of a, b, c, d,

$$2(xz + yw)^2 \leq (x^2 + y^2)(z^2 + w^2).$$

Then we have

$$2(ab + cd)^2 \leq (a^2 + c^2)(b^2 + d^2),$$

$$2(ac + db)^2 \leq (a^2 + d^2)(c^2 + b^2),$$

$$2(ad + bc)^2 \leq (a^2 + b^2)(d^2 + c^2).$$

Adding them, we get

$$2(a^2 b^2 + c^2 d^2 + a^2 c^2 + b^2 d^2 + a^2 d^2 + b^2 c^2) + 12abcd$$

$$\leq 2(a^2 b^2 + c^2 d^2 + a^2 c^2 + b^2 d^2 + a^2 d^2 + b^2 c^2).$$

Then $abcd \leq 0$, contradicting $abcd > 0$! $\qquad\square$

Solution 2 Let x, y, z, w be a permutation of a, b, c, d satisfying $x \geq z \geq y \geq w$. We will prove that it meets the required condition. Since

$$(x^2 + y^2)(z^2 + w^2) - (xz + yw)^2 = (xw - yz)^2,$$

we only need prove $(xz + yw)^2 > (xw - yz)^2$, or

$$|xz + yw| > |xw - yz|. \qquad (1)$$

From $xyzw > 0$, it is easy to see that $xz > 0$ and $yw > 0$. Note when simultaneously changing the signs of x, z or that of y, w, Formula (1) does not change, so we may assume that $x, y, z, w > 0$. Then we have

$$|xz + yw| = xz + yw > xz > \max\{xw, yz\} > |xw - yz|.$$

Therefore, $\boxed{1}$ holds. The proof is completed. $\qquad\qquad\square$

2   As shown in Fig. 2.1, suppose $\odot O_1$ and $\odot O_2$ intersect at points P and Q, and one of the two circles' external common tangent lines touches them at A and B, respectively. Circle Γ passing through A and B intersects $\odot O_1$ and $\odot O_2$ at D and C, respectively. Prove: $\dfrac{CP}{CQ} = \dfrac{DP}{DQ}$. (Contributed by Zhang Duanyang)

Solution   As shown in Fig. 2.2, according to Monge's theorem, straight lines AD, QP, BC intersect at one point defined as K. Since $\triangle KPD \backsim \triangle KAQ$, $\dfrac{DP}{AQ} = \dfrac{KP}{KA}$. Since $\triangle KPA \backsim \triangle KDQ$, $\dfrac{AP}{DQ} = \dfrac{KA}{KQ}$. So we have,

$$\frac{AP \cdot DP}{AQ \cdot DQ} = \frac{KP}{KQ}.$$

In the same way, $\dfrac{BP \cdot CP}{BQ \cdot CQ} = \dfrac{KP}{KQ}$.

Therefore,

$$\frac{AP \cdot DP}{AQ \cdot DQ} = \frac{BP \cdot CP}{BQ \cdot CQ}. \qquad\qquad \boxed{1}$$

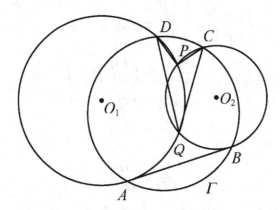

Fig. 2.1

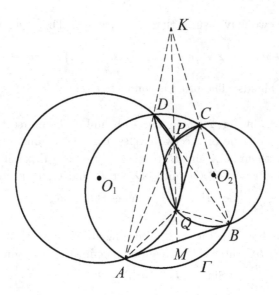

Fig. 2.2

Extend PQ to intersect with AB at point M. Since $\triangle AQM \backsim \triangle PAM$, we have

$$\frac{AQ}{AP} = \frac{AM}{PM} = \frac{QM}{AM}.$$

Then

$$\left(\frac{AQ}{AP}\right)^2 = \frac{AM}{PM} \cdot \frac{QM}{AM} = \frac{QM}{PM}.$$

In the same way, $\left(\dfrac{BQ}{BP}\right)^2 = \dfrac{QM}{PM}$. Therefore,

$$\left(\frac{AQ}{AP}\right)^2 = \left(\frac{BQ}{BP}\right)^2,$$

$$\frac{AQ}{AP} = \frac{BQ}{BP}. \qquad ②$$

From ①, ② we get

$$\frac{CP}{CQ} = \frac{DP}{DQ}. \qquad \square$$

3 Given positive integers $n, k, k \le n - 2$, and real number set $\{a_1, a_2, \ldots, a_n\}$, suppose the absolute sum of any k elements of

$\{a_1, a_2, \ldots, a_n\}$ is not greater than 1. Prove: if $|a_1| \geq 1$, then for any $2 \leq j \leq n$,

$$|a_1| + |a_i| \leq 2.$$

(Contributed by Leng Gangsong)

Solution We may assume that $a_1 \geq 1$. Then we just need prove that for any $2 \leq j \leq n$, $a_j \geq a_1 - 2$ and $a_j \leq 2 - a_1$. Define $[n] = \{1, 2, \ldots, n\}$.

We first prove $a_j \geq a_1 - 2$ for any $2 \leq j \leq n$. Select two k-element subsets I, J of $[n]$, such that $I \backslash J = \{1\}, J \backslash I = \{j\}$. Then by the given condition,

$$\sum_{s \in I} a_s \leq 1, \quad \sum_{s \in J} a_s \geq -1.$$

From the difference between these two inequalities we obtain $a_1 - a_j \leq 2$, i.e. $a_j \geq a_1 - 2$.

Next we prove $a_j \leq 2 - a_1$. Define $S = \{i \in [n] \mid a_i > 0\}$. Then $1 \in S$. If $|S| \geq k$, select a k-element subset I of S satisfying $1 \in I$. Then we have

$$0 < \sum_{s \in I \backslash \{1\}} a_s \leq 1 - a_1 \leq 0.$$

Impossible! So $|S| \leq k - 1$, and

$$|[n] \backslash (S \cup \{j\})| \geq n - k \geq 2.$$

There are then $i' \neq j' \in [n] \backslash \{1, j\}$ such that $a_{i'} \leq 0$, $a_{j'} \leq 0$. Select two k-element subsets
I and I' of $[n]$ satisfying $I \backslash I' = \{1, j\}, I' \backslash I = \{i', j'\}$. Then

$$\sum_{s \in I} a_s \leq 1, \quad \sum_{s \in I'} a_s \geq -1.$$

From the difference between these two inequalities we obtain

$$a_j + a_1 - a_{i'} - a_{j'} \leq 2.$$

Therefore

$$a_j \leq 2 - a_1 + a_{i'} + a_{j'} \leq 2 - a_1.$$

The proof is completed. □

4 Define a transformation on an integer array with n elements as

$$(a_1, a_2, \ldots, a_{n-1}, a_n) \to (a_1 + a_2, a_2 + a_3, \ldots, a_{n-1} + a_n, a_n + a_1).$$

Find all positive integer pairs $(n, k)(n, k \geq 2)$, satisfying: for any integer array with n elements $(a_1, a_2, \ldots, a_n)$, after a finite number of transformations, each number in the array obtained is a multiple of k. (Contributed by Zhang Xinze)

Solution   We assert that $(n, k) = (2^p, 2^q), p, q \in \mathbb{N}^+$. $\square$

Lemma   *Denote the array obtained after t transformations on $(a_1, a_2, \ldots, a_n)$ by $(a_1^{(t)}, a_2^{(t)}, \ldots, a_n^{(t)})$. Then (if $i = l + sn, l, s \in \mathbb{N}, 0 \leq l < n$, define $a_i = a_l$)*

$$a_i^{(t)} = a_i C_t^0 + a_{i+1} C_t^1 + \cdots + a_{i+t} C_t^t, \quad i = 1, 2, \ldots, n.$$

Proof of lemma   We will prove the lemma by mathematical induction. When $t = 1$, the statement in the lemma is obviously true. Assume that

$$a_i^{(t)} = a_i C_t^0 + a_{i+1} C_t^1 + \cdots + a_{i+t} C_t^t, \quad i = 1, 2, \ldots, n.$$

Then

$$
\begin{aligned}
a_i^{(t+1)} &= a_i^{(t)} + a_{i+1}^{(t+1)} \\
&= (a_i C_t^0 + a_{i+1} C_t^1 + \cdots + a_{i+t} C_t^t) \\
&\quad + (a_{i+1} C_t^0 + a_{i+2} C_t^1 + \cdots + a_{i+1+t} C_t^t) \\
&= a_i C_t^0 + a_{i+1}(C_t^1 + C_t^0) + a_{i+2}(C_t^2 + C_t^1) + \cdots + a_{i+t}(C_t^t + C_t^{t-1}) \\
&\quad + a_{i+1+t} C_t^t \\
&= a_i C_{t+1}^0 + a_{i+1} C_{t+1}^1 + \cdots + a_{i+t+1} C_{t+1}^{t+1}.
\end{aligned}
$$

By mathematical induction, the lemma is true for any positive integer t.

Now we assume that the required n, k exist. Note that in an array the sum of the n numbers obtained after each transformation is twice the sum of the original n numbers. Let

$$a_1 = 1, a_2 = a_3 = \cdots = a_n = 0.$$

Since after a finite number of transformations (we may assume m transformations) each number obtained is a multiple of k and the sum of these numbers is 2^m, then $k \mid 2^m$. Therefore, k is a power of 2. Furthermore, after m transformations each number obtained is a multiples of 2.

Let $2^s > m$, $s \in \mathbb{N}^+$ Note that $C_{2^s}^i = \dfrac{2^s}{i} C_{2^s-1}^{i-1} (1 \le i \le 2^s - 1)$

is an even number. $\qquad\qquad\qquad\qquad\qquad\qquad\qquad$ ①

Then after 2^s transformations, we have $a_1^{(2^s)} \equiv a_1 + a_{1+2^s} \equiv 0 \pmod 2$. So $a_{1+2^s} = 1 = a_1$, and that means $n \mid 2^s$. Therefore, n is also a power of 2.

On the other hand, we will show that, when $(n, k) = (2^p, 2^q)$, $p, q \in \mathbb{N}^+$, any integer array with n elements can be, after a finite number of transformations, changed into an array with each element being a multiple of k.

Combing the lemma and ①, after $n = 2^p$ transformations on an integer array $(a_1, a_2, \ldots, a_n)$ we have

$$a_i^{(n)} \equiv a_i + a_{i+n} \equiv 0 \pmod 2, \quad i = 1, 2, \ldots, n.$$

Then after $n = 2^p$ transformations on $\left(\dfrac{1}{2} a_1^{(n)}, \dfrac{1}{2} a_2^{(n)}, \ldots, \dfrac{1}{2} a_n^{(n)} \right)$, each number in the obtained array is also an even number. Therefore, $a_i^{(2n)} \equiv 0 \pmod 4$, $i = 1, 2, \ldots, n$.

By mathematical reduction, we get $a_i^{(qn)} \equiv 0 \pmod{2^q}$, $i = 1, 2, \ldots, n$, i.e. it is a multiple of $k = 2^q$.

The proof is completed. $\qquad\qquad\qquad\qquad\qquad\qquad\qquad$ □

Second Day
(8:00 – 12:00; August 16, 2016)

5 Prove: there are infinitely many groups of positive integers (a, b, c) that satisfy

(1) a, b, c are coprime with each; and
(2) $ab + c, bc + a, ca + b$ are coprime with each other.

(Contributed by Zhang Duanyang)

Solution Take any positive integer k so that $k - 1$ is not a multiple of 5. Then the group of positive integers $(2k-1, 2k, 2k+1)$ meet the requirements in the question.

Firstly, $2k - 1, 2k, 2k + 1$ are obviously coprime with each other.

Next, we have

$$(2k-1)2k + (2k+1) = 4k^2 + 1,$$

$$2k(2k+1) + (2k-1) = 4k^2 + 4k - 1,$$

$$(2k+1)(2k-1) + 2k = 4k^2 + 2k - 1.$$

Since $4k^2 + 1$ is odd, then

$$(4k^2 + 1, 4k^2 + 4k - 1) = (4k^2 + 1, \quad 4k - 2) = (4k^2 + 1, 2k - 1)$$

$$= (2, 2k - 1) = 1.$$

Since $k - 1$ is not a multiple of 5, then

$$(4k^2 + 1, 4k^2 + 2k - 1) = (4k^2 + 1, 2k - 2) = (4k^2 + 1, k - 1)$$

$$= (5, k - 1) = 1.$$

Finally, $(4k^2 + 4k - 1, 4k^2 + 2k - 1) = (4k^2 + 4k - 1, 2k) = 1.$

Therefore, $(2k - 1, 2k, 2k + 1)$ meet the requirements in the question, and there are infinitely many of such a kind of k. The proof is finished. $\square$

6 Let $a_1, a_2, \ldots, a_n$ be non-negative real numbers, and $S_k = \sum_{i=1}^{k} a_i$, $1 \le k \le n$. Prove:

$$\sum_{i=1}^{n} \left(a_i S_i \sum_{j=1}^{n} a_j^2 \right) \le \sum_{i=1}^{n} (a_i S_i)^2.$$

(Contributed by Wang Guangting)

Solution 1 Let $b_i = a_i S_i$, $c_i = \sum_{j=i}^{n} a_j^2$, $i = 1, 2, \ldots, n$. Then the original inequality is equivalent to

$$\sum_{i=1}^{n} b_i c_i \le \sum_{i=1}^{n} b_i^2. \qquad \qquad ①$$

Note that for $1 \le i \le n$, we have

$$B_i = b_1 + b_2 + \cdots + b_i$$

$$= a_1 S_1 + a_2 S_2 + \cdots + a_i S_i$$

$$\le (a_1 + a_2 + \cdots + a_i) S_i$$

$$= S_i^2.$$

Then by Abel's Identity,

$$\sum_{i=1}^{n} b_i c_i = \sum_{i=1}^{n-1} B_i(c_i - c_{i+1}) + B_n c_n$$

$$\leq \sum_{i=1}^{n-1} a_i^2 S_i^2 + B_n c_n$$

$$\leq \sum_{i=1}^{n} a_i^2 S_i^2 = \sum_{i=1}^{n} b_i^2.$$

Therefore, ① holds. The proof is completed. □

Solution 2 Since $\sum_{i=1}^{j} a_i S_i \leq \left(\sum_{i=1}^{j} a_i\right) S_j = S_j^2$, we have

$$\sum_{i=1}^{n} \left(a_i S_i \sum_{j=i}^{n} a_j^2\right) = \sum_{j=1}^{n} \left(a_j^2 \sum_{i=1}^{j} a_i S_i\right)$$

$$= \sum_{j=1}^{n} a_j^2 \left(\sum_{i=1}^{j} a_i S_i\right) \leq \sum_{j=1}^{n} a_j^2 S_j^2.$$

The proof is completed. □

7. As shown in Fig. 7.1, $ABCD$ is a quadrilateral inscribed in a circle, and $\angle BAC = \angle DAC$. Let $\odot I_1$, $\odot I_2$ be the inscribed circles of $\triangle ABC$ and $\triangle ADC$, respectively. Prove: one of the external common tangent lines of $\odot I_1$, $\odot I_2$ is parallel to BD. (Contributed by Yang Mingliang)

Solution As shown in Fig. 7.2, let I be the incenter of $\triangle ABD$, connecting BI. Through I draw a line tangent to $\odot I_1$ at point E, and cross AB at point M.

By chicken claw theorem, $CI = CB$. As the sums of the lengths of the opposite sides of a circumscribed quadrilateral is equal, we have $CI + MB = CB + MI$. Therefore, $MB = MI$, and $\angle MBI = \angle MIB$. Note I is the incenter of $\triangle ABD$, so $\angle MBI = \angle DBI$, and $\angle MIB = \angle DBI$. Consequently, $IE // BD$.

In the same way, pass I draw a line tangent to $\odot I_2$ at point F, and we have $IF // BD$.

Therefore, the three points E, I, F are collinear, so we get an external common tangent line EF of $\odot I_1$, $\odot I_2$ that is parallel to BD.

The proof is completed. □

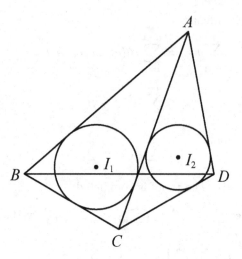

Fig. 7.1

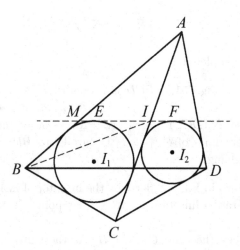

Fig. 7.2

8 Given integers $m, n, 2 \le m < n$, $(m, n) = 1$, find the smallest integer k, satisfying: for any m-element subset I of set $\{1, 2, \ldots, n\}$, if $\sum_{i \in I} i > k$, then there are n real numbers $a_1 \le a_2 \le \cdots \le a_n$, such that

$$\frac{1}{m} \sum_{i \in I} a_i > \frac{1}{n} \sum_{i=1}^{n} a_i.$$

(Contributed by Zou Jin)

Solution We will prove that the required smallest integer is $k = \dfrac{mn + m - n + 1}{2}$.

Firstly, we show $k = \dfrac{mn + m - n + 1}{2}$ meet the given condition. Let $I = \{i_1, i_2, \ldots, i_m\}$ be an m-element subset of $\{1, 2, \ldots, n\}$ with $1 \leq i_1 < i_2 < \cdots < i_m \leq n$ and satisfy $\sum_{i \in I} i > k$. As $(m, n) = 1$, we have

$$\sum_{r=1}^{m} \left[(r-1)\frac{n}{m} \right] = \frac{1}{2} \left\{ \sum_{r=1}^{m-1} \left[r \cdot \frac{n}{m} \right] + \sum_{r=1}^{m-1} \left[(m-r) \cdot \frac{n}{m} \right] \right\}$$

$$= \frac{1}{2} \cdot (n-1) \cdot (m-1). \text{ (Define } [x] \text{ as the largest integer not greater than } x)$$

Then

$$\sum_{r=1}^{m} \left\{ \left[(r-1)\frac{n}{m} \right] + 1 \right\} = k < \sum_{r=1}^{m} i_r.$$

Therefore, there exists $1 \leq r \leq m$ so that $i_r > \left[(r-1)\dfrac{n}{m} \right] + 1$. Let

$$a_1 = \cdots = a_{i_r-1} = 0, \ a_{i_r} = a_{i_r+1} = \cdots = a_n = 1.$$

Then we have

$$\frac{1}{m} \cdot \sum_{i \in I} a_i = \frac{m - r + 1}{m},$$

$$\frac{1}{n} \cdot \sum_{i=1}^{n} a_i = \frac{n - i_r + 1}{n}.$$

Since integer $i_r > \left[(r-1)\dfrac{n}{m} \right] + 1$, we have $i_r > (r-1)\dfrac{n}{m} + 1$. Therefore,

$$\frac{1}{m} \sum_{i \in I} a_i > \frac{1}{n} \sum_{i=1}^{n} a_i.$$

Next we show that $k < \dfrac{mn + m - n + 1}{2}$ cannot meet the given condition.

Let $i_r = \left[(r-1)\dfrac{n}{m} \right] + 1$, $r = 1, 2, \ldots, m$. Then

$$I = \{i_1, i_2, \ldots, i_m\} \subseteq \{1, 2, \ldots, n\},$$

and $\sum_{i \in I} i = \dfrac{mn + m - n + 1}{2} > k$. For real numbers $a_1 \le a_2 \le \cdots \le a_n$ we have

$$\sum_{i=1}^{n} a_i = \sum_{r=1}^{m-1} \left(\sum_{i=i_r}^{i_{r+1}-1} a_i \right) + \sum_{i=i_m}^{n} a_i$$

$$\ge \sum_{r=1}^{m-1} a_{i_r}(i_{r+1} - i_r) + a_{i_m}(n - i_m + 1)$$

$$= i_1(-a_{i_1}) + i_2(a_{i_1} - a_{i_2}) + \cdots + i_m(a_{i_{m-1}} - a_{i_m}) + (n+1)a_{i_m}$$

$$\ge 1 \cdot (-a_{i_1}) + \left(\frac{n}{m} + 1 \right)(a_{i_1} - a_{i_2}) + \cdots + \left((m-1)\frac{n}{m} + 1 \right)$$

$$\times (a_{i_{m-1}} - a_{i_m}) + (n+1)a_{i_m}$$

$$= \frac{n}{m}(a_{i_1} + a_{i_2} + \cdots + a_{i_m}).$$

Therefore $\dfrac{1}{n} \sum_{i=1}^{n} a_i \ge \dfrac{1}{m} \sum_{i \in I} a_i$, failing to meet the given condition.

In conclusion, the required smallest integer $k = \dfrac{mn + m - n + 1}{2}$. □

China Western Mathematical Olympiad

The 2017 China Western Mathematical Invitation (CWMI) was held in Nanchong, Sichuan from August 11th to 16th. Commissioned by the Organizing Committee of the Western China Mathematics Invitation and the Sichuan Mathematical Society, the CWMI 2017 was hosted by Nanchong Senior High School in Sichuan Province.

A total of 340 contestants (82 formal and 258 informal) from 21 teams participated in CWMI 2017, including teams from Hong Kong, Kazakhstan, Singapore, Indonesia, and the Philippines. The contest is divided into two days, 4 questions on each day, and 15 points for each question, a full score of 120 points. Yu Jiawei (108 points) of the Jiangxi representative team from Nanchang No. 2 Middle School won the individual championship, and the Sichuan representative team ranked first in total scores. Among the formal contestants, 27 won gold medals, 29 won silver medals, and 21 won bronze medals. The awarding of the informal contestants were in line with that of the formal ones: 64 won the first prize, 115 won the second prize, and 70 won the third prize.

The director of the main examination committee of CWMI 2017 is Leng Gangsong (Professor of Shanghai University). The list of members is as follows:

Xiong Bin (Professor of East China Normal University);

Ge Jun (High School Affiliated to Nanjing Normal University);

Zou Jin (Gaosi Education);

Qu Zhenhua (Associate Professor of East China Normal University);
He Yijie (East China Normal University);
Zhang Duanyang (High School Affiliated to Renmin University);
Wang Guangting (Shanghai High School);
Shi Zehui (High School Affiliated to Jilin University);
Zheng Fan (Postdoc of Princeton University);
Liu Shixiong (Zhongshan Affiliated High School of South China Normal University);
Feng Zhigang (Shanghai High School).

First Day
(8:00 – 12:00; August 13, 2017)

1 Let prime number p and positive integer n satisfy that $\prod_{k=1}^{n}(k^2+1)$ is divisible by p^2.

Prove: $p < 2n$. (Contributed by Wang Guangting)

Solution Since p is a prime, there are two situations for the divisibility.

(1) For some $k(1 \leq k \leq n)$, p^2 divides (k^2+1). Then $p^2 \leq n^2+1$, and

$$p \leq \sqrt{n^2+1} < 2n.$$

(2) For every $k(1 \leq k \leq n)$, k^2+1 is not divisible by p^2. Then for some $j, k, 1 \leq j \neq k \leq n$, $p \mid (j^2+1)$ and $p \mid (k^2+1)$. It follows $p \mid (k^2-j^2)$, or

$$p \mid (k-j)(k+j).$$

 (i) If $p \mid (k-j)$, then $p \leq k-j \leq n-1 < 2n$;
 (ii) If $p \mid (k+j)$, then $p \leq k+j \leq n+n-1 = 2n-1 < 2n$.

We conclude that $p < 2n$.   □

Remark The problem in its original form assumed that $\prod_{k=1}^{n}(k^3+1)$ is divisible by p^2, and asked the contestants to prove the same conclusion. After discussion, the contest committee has decided to replace $\prod_{k=1}^{n}(k^3+1)$ by $\prod_{k=1}^{n}(k^2+1)$, so that it is not so difficulty for contestants to solve. Because it is the first problem in the competition, the difficulty level should be easy or modest, and a problem with factors k^2+1 is more familiar to the students. In fact, this is only a minor difference, as one can use formula $k^3 + 1 = (k+1)(k^2 - k + 1)$ and the rest is similar to the current proof.

2 Let n be a positive integer such that there are positive integers x_1, $x_2, \ldots, x_n$, satisfying

$$x_1 x_2 \ldots x_n (x_1 + x_2 + \ldots + x_n) = 100n,$$

Find the largest possible value of n. (Contributed by Zou Jin)

Solution The largest possible value of n is 9702.

From equation

$$x_1 x_2 \ldots x_n (x_1 + x_2 + \ldots + x_n) = 100n,$$

it is clear that $x_1 + x_2 + \ldots + x_n \geq 1 + 1 + \ldots + 1 = n$, so we have $x_1 x_2 \ldots x_n \leq 100$. However, the product cannot be 100 because that requires $x_1 = x_2 = \ldots = x_n = 1$, which gives $x_1 x_2 \ldots x_n = 1$. Hence $x_1 x_2 \ldots x_n \leq 99$. From inequality

$$x_1 x_2 \ldots x_n = [(x_1 - 1) + 1][(x_2 - 1) + 1] \ldots [(x_n - 1) + 1]$$
$$\geq (x_1 - 1) + (x_2 - 1) + \ldots + (x_n - 1) + 1$$
$$= x_1 + x_2 + \ldots + x_n - n + 1,$$

it follows that

$$x_1 + x_2 + \ldots + x_n \leq x_1 x_2 \ldots x_n + n - 1 \leq n - 98.$$

Finally, we solve $99(n - 98) \geq 100n$, and deduce $n \leq 99 \times 98 = 9702$. By taking $x_1 = 99$, $x_2 = x_3 = \ldots = x_{9702} = 1$, we see $n = 9702$ can be achieved. □

3 As shown in Fig. 3.1, in $\triangle ABC$, point D is on side BC. Let I_1 and I_2 be the incenters of $\triangle ABD$ and $\triangle ACD$, respectively. Let O_1, O_2 be the circumcenters of $\triangle AI_1 D$ and $\triangle AI_2 D$, respectively. Let P be the intersection of lines $I_1 O_2$ and $I_2 O_1$. Prove: $PD \perp BC$. (Contributed by Zhang Duanyang)

Solution Since $O_1 A = O_1 I_1 = O_1 D$ and I_1 is the incenter of $\triangle ABD$, we infer that O_1 is the midpoint of $\overset{\frown}{AD}$ of the circumcircle of $\triangle ABD$. Extend BI_1, DI_2 to meet at J_1 as shown in Fig. 3.2.

Then J_1 is the excenter relative to the vertex B of $\triangle ABD$, and O_1 is the midpoint of $I_1 J_1$. Similarly, extend DI_1, CI_2 to meet at J_2. Then J_2 is the excenter relative to the vertex C of $\triangle ACD$, and O_2 is the midpoint of $I_2 J_2$.

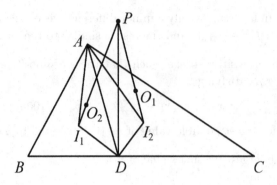

Fig. 3.1

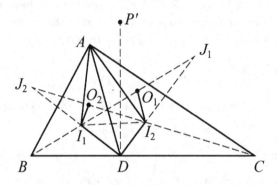

Fig. 3.2

Let line $DP' \perp BC$. It suffices to prove that I_1O_2, I_2O_1, DP' are concurrent. Applying Ceva's theorem to $\triangle DI_1I_2$, it suffices to prove

$$\frac{\sin \angle P'DI_2}{\sin \angle P'DI_1} \cdot \frac{\sin \angle DI_1O_2}{\sin \angle O_2I_1I_2} \cdot \frac{\sin \angle O_1I_2I_1}{\sin \angle DI_2O_1} = 1.$$

From $O_2J_2 = O_2I_2$, we obtain $S_{\triangle O_2I_1J_2} = S_{\triangle O_2I_1I_2}$, and

$$\frac{\sin \angle DI_1O_2}{\sin \angle O_2I_1I_2} = \frac{\sin \angle O_2I_1J_2}{\sin \angle O_2I_1I_2} = \frac{\dfrac{2S_{\triangle O_2I_1J_2}}{I_1J_2 \cdot I_1O_2}}{\dfrac{2S_{\triangle O_2I_1I_2}}{I_1I_2 \cdot I_1O_2}} = \frac{I_1I_2}{I_1J_2}.$$

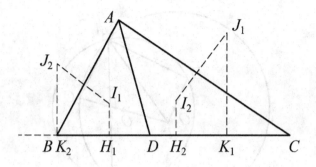

Fig. 3.3

Analogously, we obtain $\dfrac{\sin \angle O_1 I_2 I_1}{\sin \angle D I_2 O_1} = \dfrac{I_2 J_1}{I_1 I_2}$. By $\dfrac{\sin \angle P' D I_2}{\sin \angle P' D I_1} = \dfrac{\cos \angle C D I_2}{\cos \angle B D I_1}$, it suffices to prove

$$\frac{I_2 J_1 \cdot \cos \angle C D I_2}{I_1 J_2 \cdot \cos \angle B D I_1} = 1,$$

and this is equivalent to proving the projections of $I_2 J_1$ and $I_1 J_2$ on BC have equal lengths.

As shown in Fig. 3.3, let H_1, H_2, K_1, K_2 be the projection points of I_1, I_2, J_1, J_2 on BC. We have

$$H_2 K_1 = D K_1 - D H_2$$
$$= \frac{1}{2}(AB + AD - BD) - \frac{1}{2}(AD + CD - AC)$$
$$= \frac{1}{2}(AB + AC - BC),$$

and analogously, $H_1 K_2 = \dfrac{1}{2}(AB + AC - BC)$. It follows $H_2 K_1 = H_1 K_2$, completing the proof. $\qquad\square$

Remark We have an equivalent form as follows, and the proof is left to the reader.

Let $\odot O_1$ and $\odot O_2$ meet at A, B. Let two perpendicular rays emanating from A intersect $\odot O_1$, O_2 at C, D, respectively, and let $D O_1$, $C O_2$ meet at P, as shown in Fig. 3.4. Then $\angle PAD = \angle BAC$.

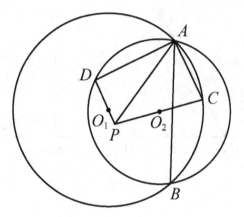

Fig. 3.4

4. Given integers n, k, $n \geq k \geq 2$, Alice and Bob play the following game on an $n \times n$ squared paper. Originally, every square on the paper is white. Alice and Bob take turns (Alice goes first) to color a white square black. If after one's turn, there is at least one black square in every $k \times k$ square, the game is over, and the player who takes the last turn wins. Which player has the winning strategy? (Contributed by Qu Zhenhua)

Solution If $n \leq 2k - 1$, Alice colors the square on the k-th row, k-th column and wins on the first move. In the following, we assume $n \geq 2k$.

We introduce the concepts of "good" and "bad" situations:

If there exist two disjoint white $k \times k$ squares on the paper and the number of white squares on the paper is odd, then we call it a good situation; if there exist two disjoint white $k \times k$ squares on the paper and the number of white squares on the paper is even, then we call it a bad situation.

Claim *If a player is facing a good situation on his turn, then the player has a winning strategy.*

Proof Suppose Alice is facing a good situation on her turn. She chooses two disjoint white $k \times k$ squares. Since the number of white squares on the paper is odd, she can find a white square elsewhere and color it black. The two disjoint $k \times k$ squares are still white, and the number of white squares on the paper becomes even.

Now Bob is facing a bad situation on his turn. If, after his move, there are still two disjoint white $k \times k$ squares, then Alice will face a good situation again. If, after his move, there are no two disjoint white $k \times k$ squares, then all the remaining white $k \times k$ squares are contained in a $2k - 1 \times 2k - 1$ squares area S. By coloring the central white square of S black, Alice can remove all the white $k \times k$ squares on the paper and wins the game.

In any case, a player who faces a good situation will either win or face a good situation again on the next turn; a player who faces a bad situation will either lose or face a bad situation again on the next turn. Since the game ends after finitely many moves, the claim is validated.

For $n \geq 2k$, there are two disjoint white $k \times k$ squares at the start of the game. If n is odd, the game starts in a good situation and Alice wins; if n is even, the game starts in a bad situation and Bob wins. $\qquad\square$

Second Day
(8:00 – 12:00; August 14, 2017)

5 Let 9 positive integers $a_1, a_2, \ldots, a_9$ (not necessarily distinct) satisfy the following condition: for any indices

$$1 \leq i < j < k \leq 9,$$

there always exists $l, 1 \leq l \leq 9$, different from i, j, k, such that $a_i + a_j + a_k + a_l = 100$. Find the number of such 9-tuples $(a_1, a_2, \ldots, a_9)$. (Contributed by He Yijie)

Solution Let $(a_1, a_2, \ldots, a_9)$ satisfy the given condition. Rename $a_1, a_2, \ldots, a_9$ in the ascending order, as $b_1 \leq b_2 \leq \cdots \leq b_9$. Clearly, there exist $l \in \{4, 5, \ldots, 9\}$ and $l' \in \{1, 2, \ldots, 6\}$, such that

$$b_1 + b_2 + b_3 + b_l = b_{l'} + b_7 + b_8 + b_9 = 100. \qquad \text{①}$$

Since

$$b_{l'} \geq b_1, b_7 \geq b_2, b_8 \geq b_3, b_9 \geq b_l, \qquad \text{②}$$

it follows from ① that all four pairs of the variables in ② must be pairwise equal, and this implies $b_2 = b_3 = \ldots = b_8$, $b_{l'} = b_1$, and $b_l = b_9$.

Thereby we may assume $(b_1, b_2, \ldots, b_9) = (x, y, \ldots, y, z)$, $x \leq y \leq z$. From ①, then we have

$$x + 2y + z = 100. \qquad \text{③}$$

(1) If $x = y = z = 25$, $(b_1, b_2, \ldots, b_9) = (25, 25, \ldots, 25)$, which gives one solution tuple $(a_1, a_2, \ldots, a_9)$.

(2) If exactly one of x, z is equal to y, let the other one be w. From ③ we have $w + 3y = 100$. This condition is also sufficient. Now y may take $1, 2, \ldots, 24, 26, 27, \ldots, 33$, totally 32 different values, and every such y corresponds to one tuple $(b_1, b_2, \ldots, b_9)$, which further corresponds to 9 tuples $(a_1, a_2, \ldots, a_9)$, altogether $32 \times 9 = 288$ solution tuples $(a_1, a_2, \ldots, a_9)$.

(3) If $x < y < z$, then by the given condition, there exists $b_l \in \{x, y, z\}$, such that

$$3y + b_l = 100.$$

Comparing it with ③, we obtain $y + b_l = x + z$, $b_l = y$, and hence $y = 25$, $x + z = 50$. It is easy to see this is also sufficient. For every $x = 1, 2, \ldots, 24$, there exists one tuple $(b_1, b_2, \ldots, b_9)$, which gives $9 \times 8 = 72$ tuples $(a_1, a_2, \ldots, a_9)$, altogether $24 \times 72 = 1728$ solution tuples $(a_1, a_2, \ldots, a_9)$.

Combining the results in (1), (2), (3), we have $1 + 288 + 1728 = 2017$ 9-tuples in total. □

6 As shown in Fig. 6.1, in acute $\triangle ABC$, points D, E are on AB, AC, respectively.

Let BE, DC meet at H, and M, N be the midpoints of BD, CE, respectively. Show that H is the orthocenter of $\triangle AMN$ if and only if B, C, E, D are concyclic and $BE \perp CD$.

(Contributed by Shi Zehui)

Solution   Extend MH to meet AC at P, and extend NH to meet AB at Q, as shown in Fig. 6.2.

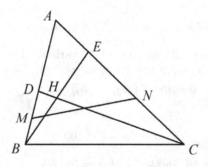

Fig. 6.1

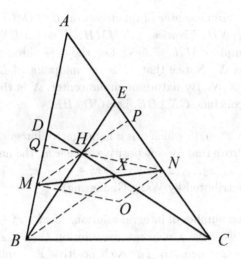

Fig. 6.2

For sufficiency. Since B, C, E, D are concyclic, $\angle BDH = \angle CEH$. It follows from $BE \perp CD$ that $DHB, \triangle EHC$ are right triangles, M, N are the midpoints of the hypotenuses BD, CE, and $\angle MDH = \angle MHD, \angle MHB = \angle MBH$. So we have

$$\angle EHP + \angle HEC = \angle MHB + \angle HDB$$

$$= \angle MBH + \angle HDB = 90°,$$

i.e., $MH \perp AC$. Analogously, $NH \perp AB$, and thus H is the orthocenter of $\triangle AMN$.

For necessity. Let H be the orthocenter of $\triangle AMN$. We have $MP \perp AN$, $NQ \perp AM$, and hence

$$\frac{DQ}{QB} = \frac{DH \cdot \sin \angle DHQ}{BH \cdot \sin \angle BHQ} = \frac{DH \cdot \sin \angle CHN}{BH \cdot \sin \angle EHN} = \frac{DH \cdot EH}{BH \cdot CH}.$$

Analogously, $\dfrac{EP}{PC} = \dfrac{DH \cdot EH}{BH \cdot CH}$, and thus $\dfrac{EP}{PC} = \dfrac{DQ}{QB}$. By nature of ratios and $DM = MB$, $EN = NC$, we get

$$\frac{EC}{PC} = \frac{DB}{QB} \Rightarrow \frac{NC}{PC} = \frac{MB}{QB} \Rightarrow \frac{NC}{PN} = \frac{MB}{QM} \Rightarrow \frac{EN}{PN} = \frac{DM}{QM}.$$

Furthermore, since H is the orthocenter of $\triangle AMN$, we have $\angle DMH = \angle ENH$, $\dfrac{QM}{MH} = \dfrac{PN}{NH}$, $\dfrac{DM}{MH} = \dfrac{EN}{NH}$, and so $\triangle DMH \backsim \triangle ENH$, $\angle MDH = \angle NEH$, which implies that B, C, E, D are concyclic.

Let O be the circumcenter of quadrilateral $BCED$. Clearly, $OM \perp AB$, and hence $OM // NH$; likewise, $ON // MH$. Hence $MHNO$ is a parallelogram, which implies $MH = ON$. Let the line through B parallel to MH meet DC at X. Notice that M is the midpoint of DB, and we have $BX = 2MH = 2ON$. By nature of circumcenter, X is the orthocenter of $\triangle BCE$ and we conclude $CX \perp BE$, i.e., $CD \perp BE$. $\square$

7 Given $n = 2^\alpha \cdot q$, in which α is a nonnegative integer, and q is an odd integer. Prove that for any positive integer m, the number of integral solutions $(x_1, x_2, \ldots, x_n)$ of $x_1^2 + x_2^2 + \cdots + x_n^2 = m$ is divisible by $2^{\alpha+1}$. (Contributed by Wang Guangting)

Solution   Let the number of integral solutions of $x_1^2 + x_2^2 + \cdots + x_n^2 = m$ be $N(m)$. Suppose we have a nonnegative solution $(x_1, x_2, \ldots, x_n)$ in which exactly k entries are nonzero. For each positive x_i, replacing it by $-x_i$ gives another solution. Since there are two choices for each nonzero entry, this particular solution gives 2^k integral solutions, and they all have the same $|x_1|, |x_2|, \ldots, |x_n|$ values. Let $S_k (k = 1, 2, \ldots, n)$ be the number of nonnegative solutions in which exactly k entries are nonzero. We have

$$N(m) = \sum_{k=1}^{n} 2^k \cdot S_k.$$

Clearly, there are C_n^k ways to choose k entries from n positions. So $C_n^k \mid S_k$. Then in order to prove $2^{\alpha+1} \mid N(m)$, it is suffice to prove $2^{\alpha-k+1} \mid C_n^k$.

Notice that $C_n^k = \dfrac{n(n-1)\ldots(n-k+1)}{k!}$, the numerator has at least α of 2's, while the number of 2's in the denominator is $\sum_{i=1}^{[\log_2 k]} \left[\dfrac{k}{2^i}\right] < \sum_{i=1}^{\infty} \dfrac{k}{2^i} = k$, which is at most $k - 1$. It follows $2^{\alpha-k+1} \mid C_n^k$, and $2^{\alpha+1} \mid N(m)$. The proof is complete.

Remark   To prove $2^{\alpha-k+1} \mid C_n^k$, one can also apply Kummer's theorem: for positive integers $n, i, i \leq n$, and p prime, $p^t \parallel C_n^i$ if and only if in the base-p system, there are at most $t (t \geq 0)$ carries in the process of adding $(n - i)$ to i.

Solution 2   Denote $f(n, m)$ as the number of integral solutions of $x_1^2 + x_2^2 + \ldots + x_n^2 = m$. We assert the following recurrence formula for $f(n, m)$.

Lemma $f(2n, m) = 2f(n, m) + \sum_{k=1}^{m-1} f(n, k)f(n, m - k)$.

Proof of the lemma   Let$(x_1, x_2, \ldots, x_{2n})$ be a solution of

$$x_1^2 + x_2^2 + \cdots + x_{2n}^2 = m,$$

for which

$$x_1^2 + x_2^2 + \cdots + x_n^2 = k.$$

If $k = 0$, $(x_1, x_2, \ldots, x_n) = (0, 0, \ldots, 0)$, and $x_{n+1}^2 + x_{n+2}^2 + \cdots + x_{2n}^2 = m$. By definition, there are $f(n, m)$ of integral solutions $(x_{n+1}, x_{n+2}, \ldots, x_{2n})$. Hence there are $f(n, m)$ solutions $(x_1, x_2, \ldots, x_{2n})$ when $k = 0$.

Clearly, there are $f(n, m)$ solutions $(x_1, x_2, \ldots, x_{2n})$ as well when $k = m$.

If $1 \leq k \leq m - 1$ and k is an integer, equation $x_1^2 + x_2^2 + \cdots + x_n^2 = k$ has $f(n, k)$ solutions, while the equation $x_{n+1}^2 + x_{n+2}^2 + \cdots + x_{2n}^2 = m - k$ has $f(n, m - k)$ solutions. It follows that $x_1^2 + x_2^2 + \cdots + x_{2n}^2 = m$ has $f(n, k)f(n, m - k)$ solutions. So we have

$$f(2n, m) = 2f(n, m) + \sum_{k=1}^{m-1} f(n, k)f(n, m - k).$$

Now we prove the original statement: when $n = 2^\alpha \cdot q$,

$$2^{\alpha+1} \mid f(n, m). \qquad \qquad \text{①}$$

Use induction on α.

If $\alpha = 0$, $(x_1, x_2, \ldots, x_n)$ and $(-x_1, -x_2, \ldots, -x_n)$ are a pair of solutions; since $m > 0$, they are distinct. It follows that all solutions can be made into pairs and hence

$$2 = 2^{0+1} \mid f(n, m).$$

Assuming that ① holds for α, we consider $\alpha + 1$. Notice

$$f(n, m) = 2f\left(\frac{n}{2}, m\right) + \sum_{k=1}^{m-1} f\left(\frac{n}{2}, k\right) f\left(\frac{n}{2}, m - k\right).$$

Since $n = 2^{\alpha+1} \cdot q$, $\frac{n}{2} = 2^\alpha \cdot q$, by the induction hypothesis, $2^{\alpha+1}$ divides $f\left(\frac{n}{2}, m\right)$, $f\left(\frac{n}{2}, k\right)$, and $f\left(\frac{n}{2}, m - k\right)$. Thus

$$2^{\alpha+2} \mid 2f\left(\frac{n}{2}, m\right), 2^{2(\alpha+1)} \mid \sum_{k=1}^{m-1} f\left(\frac{n}{2}, k\right) f\left(\frac{n}{2}, m - k\right).$$

From $2(\alpha + 1) \geq \alpha + 2$, we obtain $2^{\alpha+2} \mid f(n, m)$, completing the induction and the proof.   □

8 Let $n \geq 2$ be an integer. Show that for any positive real numbers $a_1, a_2, \ldots, a_n$, the following inequality holds:

$$\sum_{i=1}^{n} \max\{a_1, a_2, \ldots, a_i\} \cdot \min\{a_i, a_{i+1}, \ldots, a_n\} \leq \frac{n}{2\sqrt{n-1}} \cdot \sum_{i=1}^{n} a_i^2.$$
$$(*)$$

(Contributed by Zhang Duanyang)

Solution Apply strong induction to integer n.

When $n = 2$, the left side of (*) $= a_1 \cdot \min\{a_1, a_2\} + \max\{a_1, a_2\} \cdot a_2$.

If $a_1 \geq a_2$, (*) is equivalent to $2a_1 a_2 \leq a_1^2 + a_2^2$, which holds.

If $a_1 \leq a_2$, (*) is equivalent to $a_1^2 + a_2^2 \leq a_1^2 + a_2^2$, which holds, too.

Assume that the inequality holds for every integer between 2 and $n-1$.

Consider n. For $2 \leq i \leq n$, let $c_i = \dfrac{i}{2\sqrt{i-1}}$, and $c_1 = 1$. It is easy to check

$$c_1 = c_2 < c_3 < \ldots < c_n.$$

Let $M = \max\{a_1, a_2, \ldots, a_n\}$, and suppose $a_k = M$.

If $k = 1$, the left side of (*) $= M \sum_{i=1}^{n} \min\{a_i, a_{i+1}, \ldots, a_n\}$. From

$$\min\{a_1, a_2, \ldots, a_n\} = \min\{a_2, \ldots, a_n\} \leq \frac{1}{n-1} \sum_{i=2}^{n} a_i,$$

and for $2 \leq i \leq n$, $\min\{a_i, a_{i+1}, \ldots, a_n\} \leq a_i$, we obtain

$$\sum_{i=1}^{n} \min\{a_i, a_{i+1}, \ldots, a_n\} \leq \frac{1}{n-1} \sum_{i=2}^{n} a_i + \sum_{i=2}^{n} a_i = \frac{n}{n-1} \sum_{i=2}^{n} a_i.$$

By the average value inequality,

$$\text{the left side of } (*) \leq \frac{n}{n-1} M \sum_{i=2}^{n} a_i \leq \frac{n}{2\sqrt{n-1}} \left[M^2 + \frac{1}{n-1} \left(\sum_{i=2}^{n} a_i \right)^2 \right]$$

$$\leq \frac{n}{2\sqrt{n-1}} \left(M^2 + \sum_{i=2}^{n} a_i^2 \right) = \frac{n}{2\sqrt{n-1}} \sum_{i=1}^{n} a_i^2.$$

If $k = n$, $\min\{a_i, a_{i+1}, \ldots, a_n\} = \min\{a_i, a_{i+1}, \ldots, a_{n-1}\}$, and the left side of (*) $= \sum_{i=1}^{n-1} \max\{a_1, a_2, \ldots, a_i\} \cdot \min\{a_i, a_{i+1}, \ldots, a_{n-1}\} + M^2$.

By the induction hypothesis,

$$\sum_{i=1}^{n-1} \max\{a_1, a_2, \ldots, a_i\} \cdot \min\{a_i, a_{i+1}, \ldots, a_{n-1}\} \le c_{n-1} \sum_{i=1}^{n-1} a_i^2,$$

and thus

$$\text{the left side of } (*) \le c_{n-1} \sum_{i=1}^{n-1} a_i^2 + M^2 < \frac{n}{2\sqrt{n-1}} \left(\sum_{i=1}^{n-1} a_i^2 + M^2 \right)$$

$$= \frac{n}{2\sqrt{n-1}} \sum_{i=1}^{n} a_i^2.$$

Finally, if $2 \le k \le n-1$, by the above argument for $k = 1$ and $k = n$, we have

$$\text{the left side of } (*) = \sum_{i=1}^{k-1} \max\{a_1, a_2, \ldots, a_i\} \cdot \min\{a_i, a_{i+1}, \ldots, a_n\}$$

$$+ M \sum_{i=k}^{n} \min\{a_i, a_{i+1}, \ldots, a_n\}$$

$$\le \sum_{i=1}^{k-1} \max\{a_1, a_2, \ldots, a_i\} \cdot \min\{a_i, a_{i+1}, \ldots, a_{k-1}\}$$

$$+ \frac{n-k+1}{n-k} M \sum_{i=k+1}^{n} a_i$$

$$\le c_{k-1} \sum_{i=1}^{k-1} a_i^2 + \frac{n-k+1}{2\sqrt{n-k}} \left(M^2 + \sum_{i=k+1}^{n} a_i^2 \right)$$

$$= c_{k-1} \sum_{i=1}^{k-1} a_i^2 + c_{n-k+1} \sum_{i=k}^{n} a_i^2 < \frac{n}{2\sqrt{n-1}} \sum_{i=1}^{n} a_i^2.$$

The desired result is yielded. □

Remark Taking $a_1 = \sqrt{n-1}$, $a_2 = \ldots = a_n = 1$, one can see that the coefficient $\dfrac{n}{2\sqrt{n-1}}$ is optimal.

China Southeastern Mathematical Olympiad

2016 (Nanchang, Jiangxi)

From July 28th to August 2nd, 2016, the 13th China Southeastern Mathematical Olympiad (CSMO) was held in Liantang No.1 Middle School in Nanchang, Jiangxi Province, China. A total of 184 teams with 1,226 contestants participated in this grand event. They are from 125 famous schools at home and abroad, including Jiangxi, Fujian, Zhejiang, Beijing, Shanghai, Guangdong, Jiangsu, Shandong, Shanxi, Hebei, Shaanxi, Anhui, Gansu, Sichuan, Liaoning, Jilin, Heilongjiang, Hong Kong Special Administrative Region, Macao Special Administrative Region, Taiwan, as well as Thailand, Malaysia, and the Philippines.

The purpose of this activity is to strengthen the exchange and cooperation of mathematics competitions in Fujian, Zhejiang, and Jiangxi regions, and to promote and improve the level of mathematics competitions among middle school students in the three provinces. At the initiative of Mr. Qiu Zonghu, Research Fellower at the Chinese Academy of Sciences and former vice chairman of the Chinese Mathematical Olympic Committee, the Mathematical Societies of the three provinces formed the Mathematical Olympiad Competition Committee in Southeast China, which is sponsored by the relevant schools of the three provinces in turn. Participants are the first year high school students from the three provinces. At the same time, other first year high school students at home and abroad are invited to participate. From 2014, some second year high school students are also invited to participate, so as to provide them with the opportunity to participate in independent admissions of famous universities in China, and have face-to-face communication with admissions officers during the event.

Since the first CSMO was held in 2004, the influence of this competition has continued to expand. Key middle schools at Shanghai, Guangdong, Jiangsu, Shandong, Beijing, Hebei, Hunan, Anhui, Shaanxi, Hubei, Henan, Liaoning, Jilin, Sichuan and other places, as well as schools at Hong Kong, Macao, Taiwan regions and some overseas schools have also sent teams to participate. This event has cultivated many high-level Mathematical Olympiad contestants for the three provinces of Fujian, Zhejiang, and Jiangxi, and not a few of them have been selected into the national training team and represented the national team to gold medals and silver medals at the International Mathematical Olympiad.

The director of the current test committee is Professor Wu Quanshui of Fudan University.

The members of the committee (in alphabetical order):

Chen Zongxuan (South China Normal University);

Dong Qiuxian (Nanchang University);

He Yijie (East China Normal University);

Li Shenghong (Zhejiang University);

Sun Wenxian (Nine Chapter Mathematical Education Foundation);

Tao Pingsheng (Jiangxi Normal University of Science and Technology);

Wu Genxiu (Jiangxi Normal University);

Wu Quanshui (Fudan University);

Xiong Bin (East China Normal University);

Zhang Pengcheng (Fujian Normal University).

10th Grade
First Day
(8:00 – 12:00; July 30, 2016)

1 Suppose sequence $\{a_n\}$ satisfies $a_1 = 1$, $a_2 = \frac{1}{2}$, and for any integer $n \geq 2$,

$$n(n+1)a_{n+1}a_n + na_n a_{n-1} = (n+1)^2 a_{n+1}a_{n-1}.$$

(1) Find the general term of $\{a_n\}$;

(2) for any integer $n > 2$, prove: $\dfrac{2}{n+1} < \sqrt[n]{a_n} < \dfrac{1}{\sqrt{n}}$.

Solution (1) From $n(n+1)a_{n+1}a_n + na_na_{n-1} = (n+1)^2 a_{n+1}a_{n-1}$ we have

$$na_na_{n-1} = (n+1)a_{n+1} \cdot [(n+1)a_{n-1} - na_n],$$

or equivalently,

$$\frac{1}{(n+1)a_{n+1}} - \frac{1}{a_n} = \frac{1}{na_n} - \frac{1}{a_{n-1}}. \qquad \text{(1)}$$

Note that $a_1 = 1$, $a_2 = \dfrac{1}{2}$. Then by (1),

$$\frac{1}{na_n} - \frac{1}{a_{n-1}} = \frac{1}{(n-1)a_{n-1}} - \frac{1}{a_{n-2}} = \cdots = \frac{1}{2a_2} - \frac{1}{a_1} = 0.$$

Consequently, $a_n = \dfrac{1}{n}a_{n-1} = \dfrac{1}{n} \cdot \dfrac{1}{n-1}a_{n-2} = \cdots = \dfrac{1}{n!}.$

(2) We just need prove equivalently,

$$\sqrt{n} < \sqrt[n]{n!} < \frac{n+1}{2}. \qquad \text{(2)}$$

First, by the Mean Inequality we have

$$\sqrt[n]{n!} = \sqrt[n]{1 \cdot 2 \cdots \cdot n} < \frac{1 + 2 + \cdots + n}{n} = \frac{n+1}{2}.$$

Next, for any integer $k(1 < k < n)$, we have

$$k(n-k+1) - n = (k-1)(n-k) > 0,$$

that is $k(n-k+1) > n$. Therefore,

$$(n!)^2 = (1 \cdot n) \cdot (2 \cdot (n-1)) \cdots (k \cdot (n-k+1)) \cdots (n \cdot 1) > n^n.$$

So $\sqrt{n} < \sqrt[n]{n!}$. Inequalities (2) is then proved.

Remark In solving Question (1), we can first assume that $a_n = \dfrac{1}{n!}$, and then prove it by the Second Principle of Mathematical Induction.

2 As shown in Fig. 2.1, PAB and PCD are two secant lines of circle O, AD and BC intersect at point Q, T is a point on line BQ, line segment PT and circle O intersect at point K, and line QK and line PA intersect at point S.

Prove: If $ST \parallel PQ$, then B, S, K, and T are concyclic.

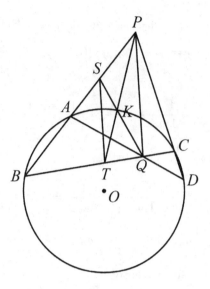

Fig. 2.1

Solution   As shown in Fig. 2.2, let segment PQ and its extension intersect with circle O at E and F, respectively. From

$$\angle PQA > \angle ADC = \angle ABC$$

we know there exists point G on PQ such that $\angle ABG = \angle PQA$. Then A, B, G, Q are four concyclic points. Therefore,

$$PQ \cdot PG = PA \cdot PB = PE \cdot PF,$$

and

$$\angle PGB = 180° - \angle BAD = 180° - \angle BCD = \angle PCB.$$

Then B, G, C, P are also four concyclic points, and

$$PQ \cdot QG = QC \cdot QB = QE \cdot QF.$$

Therefore,

$$PQ^2 = PQ \cdot PG - PQ \cdot QG = PE \cdot PF - QE \cdot QF.$$

Extend BK to intersect with PQ at point R. In $\triangle PBQ$, from Ceva's theorem we have

$$\frac{PR}{RQ} \cdot \frac{QT}{TB} \cdot \frac{BS}{SP} = 1. \tag{1}$$

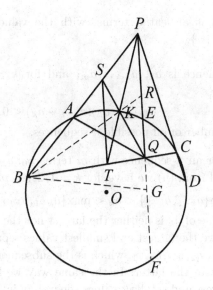

Fig. 2.2

Since $ST // PQ$, we have

$$\frac{QT}{TB} = \frac{PS}{SB}. \qquad \text{②}$$

So R is the midpoint of segment PQ. Then

$$PQ = 2RQ, PE = RQ + RE, PF = RQ + RF,$$

$$QE = RQ - RE, QF = RF - RQ,$$

Furthermore,

$$4RQ^2 = (RQ + RE)(RQ + RF) - (RQ - RE)(RF - RQ)$$

$$= 2RQ^2 + 2RE \cdot RF,$$

and that is

$$RQ^2 = RE \cdot RF \cdot RE \cdot RF = RK \cdot R.$$

Therefore, $RQ^2 = RK \cdot RB$ and $\triangle RKQ \backsim \triangle RQB$.

Consequently, $\angle KBT = \angle KQR = \angle KST$, so B, S, K, T are concyclic. □

3 Given integer $n \geq 3$ and a finite sequence, we call it an "n-sequence", if

    (i) it contains at leat 3 terms, with the value of each term in $\{1, 2, \ldots, n\}$;

    (ii) suppose

        the sequence is $a_1, a_2, \ldots, a_m$, and for $k = 1, 2, \ldots, m - 2$, it satisfies

$$(a_{k+1} - a_k)(a_{k+2} - a_k) < 0.$$

Find the number of all possible n-sequences.

Solution Consider an n-sequence with m terms S_0: $a_1, a_2, \ldots, a_m$. Then from (i) $m \geq 3$, and from (ii) we have, for $k = 1, 2, \ldots, m - 2$,

$$\min\{a_{k+1}, a_{k+2}\} < a_k < \max\{a_{k+1}, a_{k+2}\}, \qquad \textcircled{1}$$

that means the value of a_k is neither the largest nor the smallest in S_0. So a_m, a_{m-1} must have the largest and smallest values separately in S_0.

    Consider S_1: $a_1, a_2, \ldots, a_{m-1}$, which is the subsequence of S_0 obtained by dropping a_m from the latter. In the same way we know a_{m-1}, a_{m-2} must have the largest and smallest values separately in S_1. And so on, we know for $t = 0, 1, \ldots, m - 2$, in the subsequence S_t: $a_1, a_2, \ldots, a_{m-t}$ of S_0, a_{m-t}, a_{m-t-1} always have the largest and smallest values separately.

    Therefore, we have two possible cases as the following.

    Case 1: a_m has the largest value and a_{m-1} has the smallest value in S_0. Then a_{m-2} must have the largest value in S_1. And go in this way, we get finally,

$$a_m > a_{m-2} > \cdots > a_1 > a_2 > a_4 > \cdots > a_{m-1} \ (m \text{ is an odd number});$$
$$\textcircled{2}$$

$$a_m > a_{m-2} > \cdots > a_2 > a_1 > a_3 > \cdots > a_{m-1} \ (m \text{ is an even number});$$
$$\textcircled{3}$$

    Case 2: a_m has the smallest value and a_{m-1} has the largest value in S_0. Then a_{m-2} must have the smallest value in S_1. And go in this way, we get finally,

$$a_m < a_{m-2} < \cdots < a_1 < a_2 < a_4 < \cdots < a_{m-1} \ (m \text{ is an odd number});$$
$$\textcircled{4}$$

$$a_m < a_{m-2} < \cdots < a_2 < a_1 < a_3 < \cdots < a_{m-1} \ (m \text{ is an even number}).$$
$$\textcircled{5}$$

    On the other hand, when S_0 satisfies any one of Formulas $\textcircled{2}$, $\textcircled{3}$, $\textcircled{4}$, $\textcircled{5}$, then from $\textcircled{1}$ we know S_0 meets Condition (ii) in the problem.

Therefore, every m-element subset of $\{1, 2, \ldots, n\}$ can be arranged into exactly two n-sequences: one satisfies ② or ③, while the other satisfies ④ or ⑤.

Note that there are totally $2^n - C_n^0 - C_n^1 - C_n^2$ subsets of $\{1, 2, \ldots, n\}$, each containing not less than 3 elements. Therefore, the number of n-sequences is

$$2(2^n - C_n^0 - C_n^1 - C_n^2) = 2^{n+1} - n^2 - n - 2. \qquad \square$$

④ Given 4 points on the plane, they can be used as vertexes to form 4 triangles and as endpoints to formed 6 segments. If the areas of the 4 triangles are 4 different positive integers, and the lengths of the 6 segments are 6 different positive integers, we say the convex hull of these four points is a "lotus diagram".

(1) Give a specific "lotus diagram", including the areas of the corresponding 4 triangles and the lengths of corresponding 6 segments.

(2) Prove that there are infinitely many "lotus diagrams" on the plane, which are not similar to each other.

Note For a set S consisting of a finite number of points that are not collinear on the plane, its convex hull is a convex polygon area Ω (including the boundary and the interior): each vertex of Ω is a point in S, and every point in S belongs to Ω.

Solution (1) The two complete quadrilaterals shown in Fig. 4.1 and Fig. 4.2 are formed by concatenating four right-angled triangles whose sides are integers. The six sides of the obtained quadrilateral are positive integers different from each other, and the values of the areas are also mutually different positive integers, so these two quadrilaterals are lotus diagrams.

(2) Take any two Pythagorean triples (a, b, c) and (x, y, z), with

$$a^2 + b^2 = c^2, \quad x^2 + y^2 = z^2.$$

Make convex quadrilateral $ABCD$ diagonally perpendicular to each other at point O, such that

$$OA = ay, \quad OB = ax, \quad OC = bx, \quad OD = by.$$

From the Pythagorean theorem we get $AB = az$, $BC = cx, CD = bz, DA = cy$, so the distance between any two points in A, B, C, and D is a positive integer. It is easy to know that there must be an even number

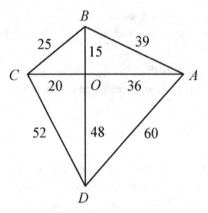

Fig. 4.1

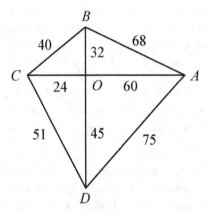

Fig. 4.2

in a and b, and an even number in x and y, so the areas of the four right-angled triangles $\triangle AOB$, $\triangle BOC$, $\triangle COD$, $\triangle DOA$ are all integers; then the areas of $\triangle ABC$, $\triangle BCD$, $\triangle CDA$ and $\triangle DAB$ must also be positive integers. Denote such convex quadrilateral $ABCD$ by $Q(a, b, c, x, y, z)$.

For any integer $t > 1$, note that $(2t, t^2 - 1, t^2 + 1)$ is a Pythagorean triple. Let

$$Q_t = Q(3, 4, 5, 2t, t^2 - 1, t^2 + 1) = A_t B_t C_t D_t.$$

We prove that integer $t > 1$ can be assigned infinitely many values, to make the convex hull of Q_t a lotus diagram. Let

$$\begin{cases} f_1(t) = A_t B_t = 3t^2 + 3, \\ f_2(t) = B_t C_t = 10t, \\ f_3(t) = C_t D_t = 4t^2 + 4, \\ f_4(t) = D_t A_t = 5t^2 - 5, \\ f_5(t) = A_t C_t = 3t^2 + 8t - 3, \\ f_6(t) = B_t D_t = 4t^2 + 6t - 4, \end{cases}$$

$$\begin{cases} g_1(t) = S_{\triangle A_t B_t C_t} = 3t(3t^2 + 8t - 3), \\ g_2(t) = S_{\triangle B_t C_t D_t} = 4t(4t^2 + 6t - 4), \\ g_3(t) = S_{\triangle C_t D_t A_t} = (2t^2 - 2)(3t^2 + 8t - 3), \\ g_4(t) = S_{\triangle D_t A_t B_t} = (3t^2 - 3)(2t^2 + 3t - 2). \end{cases}$$

For any (i, j) $(1 \le i < j \le 6)$, equation $f_i(t) = f_j(t)$ has at most two solutions in $\mathbb{N}^+$; and for any $(i, j)(1 \le i < j \le 4)$, equation $g_i(t) = g_j(t)$ also has at most two solutions in $\mathbb{N}^+$. Then there exists an infinite set $T \subseteq \mathbb{N} \backslash \{1\}$, such that for every $t \in T$, $f_k(t)(k = 1, 2, \ldots, 6)$ are different from each other, and $g_l(t)(l = 1, 2, 3, 4)$ are also different from each other. At this time, $Q_t = A_t B_t C_t D_t$ becomes a lotus diagram.

Furthermore, when $t > 3$, it is easy to find that the ratio of the longest side to the shortest side of Q_t is

$$\varphi(t) = \frac{D_t A_t}{B_t C_t} = \frac{5t^2 - 5}{10t} = \frac{1}{2}\left(t - \frac{1}{t}\right),$$

By the monotonicity of $\varphi(t)$, we know that for any $t_1, t_2 \in T$, $3 < t_1 < t_2$, $\varphi(t_1) < \varphi(t_2)$. So Q_{t_1} is not similar to Q_{t_2}. Therefore, there are infinitely many of lotus diagrams that are not similar to each other.

The proof is completed $\square$

Second Day
(8:00 – 12:00; July 31, 2016)

5　 For any positive integer n, define $f(n) = \sum_{d \in D_n} \dfrac{1}{1 + d}$, where D_n is the set of all positive factors of n. Prove: for any positive integer m,
$$f(1) + f(2) + \cdots + f(m) < m.$$

Solution   For any positive integer m, note there are exactly $\left[\dfrac{m}{d}\right]$ integers between $1, 2, \ldots, m$ which have positive integer d as their factor. So we have

$$\sum_{k=1}^{m} f(k) = \sum_{k=1}^{m} \sum_{d \mid k} \frac{1}{1+d} = \sum_{d=1}^{m} \sum_{\substack{k=1 \\ d \mid k}}^{m} \frac{1}{1+d} = \sum_{d=1}^{m} \left[\frac{m}{d}\right] \frac{1}{1+d}$$

$$\leq \sum_{d=1}^{m} \frac{m}{d} \cdot \frac{1}{1+d} = m \cdot \sum_{d=1}^{m} \left(\frac{1}{d} - \frac{1}{1+d}\right) = m \left(1 - \frac{1}{1+m}\right) < m.$$

6 Tossing a coin n (a positive integer) times independently, let $a(n)$ represents the total number of cases where the number of times that the coin's head side is up on the ground is a multiple of 3, and $b(n)$ represents the total number of cases where the number of times that the coin's head side is up is a multiple of 6.

(1) Find $a(2016)$, $b(2016)$;

(2) for $n \leq 2016$, find the number of integer n satisfying $2b(n) - a(n) > 0$.

Solution   We have

$$a(n) = C_n^0 + C_n^3 + \cdots + C_n^{3k} + \cdots,$$

$$b(n) = C_n^0 + C_n^6 + \cdots + C_n^{6k} + \cdots.$$

Let the three roots of $z^3 = 1$ be $z_k = e^{\frac{2k\pi}{3}i}(k = 1, 2, 3)$. Then for any positive integer n,

$$\sum_{k=1}^{3} z_k^n = \begin{cases} 3, & n \equiv 0 \ (\mathrm{mod}\, 3), \\ 0, & n \not\equiv 0 \ (\mathrm{mod}\, 3), \end{cases}$$

Therefore,

$$\sum_{k=1}^{3}(1+z_k)^n = \sum_{k=1}^{3}\sum_{j=0}^{n} C_n^j z_k^j = \sum_{j=0}^{n} C_n^j \left(\sum_{k=1}^{3} z_k^j\right) = \sum_{\substack{j=0 \\ 3\mid j}}^{n} 3C_n^j = 3 \cdot a(n).$$

As $1 + z_1 = e^{\frac{\pi}{3}i}$, $1 + z_2 = e^{-\frac{\pi}{3}i}$, $1 + z_3 = 2$, we have

$$a(n) = \frac{1}{3}\sum_{k=1}^{3}(1+z_k)^n = \frac{1}{3} \cdot \left(e^{\frac{n\pi}{3}i} + e^{-\frac{n\pi}{3}i} + 2^n\right)$$

$$= \frac{1}{3} \cdot \left(2^n + 2\cos\frac{n\pi}{3}\right). \tag{1}$$

In the same way, let the 6 roots of $z^6 = 1$ is $w_k = e^{\frac{k}{3}\pi i}(k = 1, 2, \ldots, 6)$. Then

$$\sum_{k=1}^{6} w_k^n = \begin{cases} 6, & n \equiv 0 \ (\text{mod}\, 6), \\ 0, & n \not\equiv 0 \ (\text{mod}\, 6), \end{cases}$$

and $\sum_{k=1}^{6}(1 + w_k)^n = 6 \cdot b(n)$. Note that $1 + w_1 = \sqrt{3}e^{\frac{\pi}{6}i}$, $1 + w_2 = e^{\frac{\pi}{3}i}$, $1 + w_3 = 0$, $1 + w_4 = e^{-\frac{\pi}{3}i}$, $1 + w_5 = \sqrt{3}e^{-\frac{\pi}{6}i}$, $1 + w_6 = 2$. We get

$$b(n) = \frac{1}{6}\sum_{k=1}^{6}(1 + w_k)^n = \frac{1}{6} \cdot \left(2^n + 2(\sqrt{3})^n \cos\frac{n\pi}{6} + 2\cos\frac{n\pi}{3}\right). \quad \text{②}$$

(1) When $n = 2016$, from ① and ② we have

$$a(2016) = \frac{1}{3} \cdot (2^{2016} + 2);$$

$$b(2016) = \frac{1}{6} \cdot (2^{2016} + 2(\sqrt{3})^{2016} + 2) = \frac{1}{6} \cdot (2^{2016} + 2 \cdot 3^{1008} + 2).$$

(2) We have

$$\begin{aligned} 2b(n) - a(n) &= \frac{1}{3} \cdot \left(2^n + 2(\sqrt{3})^n \cos\frac{n\pi}{6} + 2\cos\frac{n\pi}{3}\right) \\ &\quad - \frac{1}{3} \cdot \left(2^n + 2\cos\frac{n\pi}{3}\right) \\ &= \frac{2(\sqrt{3})^n}{3}\cos\frac{n\pi}{6}. \end{aligned}$$

For positive integer n, $2b(n) - a(n) > 0$ if and only if $\cos\dfrac{n\pi}{6} > 0$, i.e.,

$$n \equiv 0, \pm 1, \pm 2 \ (\text{mod}\, 12).$$

Therefore, in a complete residue system of module 12, exactly 5 numbers meet the conditions. So there are $\dfrac{5}{12} \times 2016 = 840$ positive integers n satisfying $2b(n) - a(n) > 0$.

7 As shown in Fig. 7.1, point I is the incenter of $\triangle ABC$, whose sides BC, CA, and AB is tangent to the inscribed circle at points D, E, F, respectively; lines BI, CI, DI intersect with EF at points M, N, K, respectively; lines BN and CM intersect at point P, lines AK and BC intersect at point G; the line through I perpendicular to PG and the line through P perpendicular to PB intersect at point Q. Prove: line BI bisects line segment PQ.

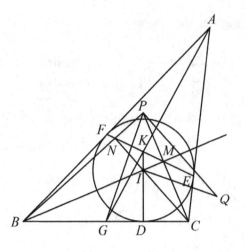

Fig. 7.1

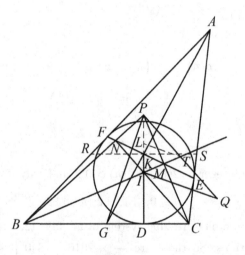

Fig. 7.2

Solution　As shown in Fig. 7.2, through K draw line parallel to BC intersecting with AB and AC at points R and S, respectively. Since $IK \perp RS$ and $IF \perp AB$, $\angle RKI = \angle RFI = 90°$, then I, K, S, E are four concyclic points. In the same way, I, K, S, E are also four concyclic points. Then

$$\angle IRK = \angle IFK = \angle IEK = \angle ISK.$$

Therefore, $KR = KS$. As $RS // BC$, G is the midpoint of BC. Since

$$\angle CNE = \angle AEF - \angle ECN = 90° - \frac{\angle BAC}{2} - \frac{\angle ACB}{2}$$

$$= \frac{\angle ABC}{2} = \angle FBI,$$

B, I, N, F are then four concyclic points. It is easy to see that B, D, I, F are also four concyclic points. Therefore, B, D, I, N, F are five concyclic points. Then $IN \perp BN$. In the same way, $IM \perp CM$. So I is the orthocenter of $\triangle PBC$, and $PI \perp BC$, i.e., P, I, D are collinear.

Suppose BI and PQ intersect at T. From $\angle PIM = \angle PCB$, we know D, C, M, I are four concyclic points. Moreover,

$$\angle TPI = 90° - \angle BPD = \angle PBC.$$

Therefore, $\triangle TPI \backsim \triangle PBC$. Let L be the midpoint of PI. $TPL \backsim PBG$, and

$$\angle PTL + \angle GPT = \angle BPG + \angle GPT = \angle BPQ = 90°.$$

Therefore, $TL // QI$, and then $PT = TQ$, which means BI bisects PQ. □

---

8 Suppose positive integer sequence $\{a_n\}$ satisfies: $n^2 \mid (a_1 + a_2 + \cdots + a_n)$ and $a_n \leq (n + 2016)^2$, for any integer $n \geq 2016$. Let $b_n = a_{n+1} - a_n$. Prove: $\{b_n\}$ is a constant sequence after a certain terms.

Solution Let $c_n = \dfrac{a_1 + a_2 + \cdots + a_n}{n^2} \ (n = 1, 2, \ldots)$. Then by the given condtion, for any $n \geq 2016$, c_n is an integer, and

$$c_n = \frac{a_1 + \cdots + a_{n-1}}{n^2} + \frac{a_n}{n^2} = \frac{(n-1)^2}{n^2} c_{n-1} + \frac{a_n}{n^2}. \qquad \textcircled{1}$$

We will prove that $\{c_n\}$ is a constant sequence after a certain term.

First, letting $M = \max\{c_1, c_2, \ldots, c_{2016}, 10\,000\}$, we will prove by mathematical induction that M is an upper bound of $\{c_n\}$. When $k = 2016$, certainly $c_k \leq M$. Assume $c_{k-1} \leq M$ for $k > 2016$. By $\textcircled{1}$ we have

$$c_k = \left(1 - \frac{2k-1}{k^2}\right) c_{k-1} + \frac{a_k}{k^2} \leq \left(1 - \frac{2k-1}{k^2}\right) M + \frac{(k+2016)^2}{k^2}$$

$$= M + 1 - \frac{1}{k^2}[(2k-1)M - 2016(2k + 2016)] < M + 1. \qquad \textcircled{2}$$

The last inequality in $\textcircled{2}$ holds because

$$(2k-1)M - 2016(2k + 2016) > k \cdot 10000 - 2016 \cdot 3k > 0.$$

Since c_k is an integer, $c_k \leq M$. Therefore, M is an upper bound of $\{c_n\}$.

When integer $n > 2M = t_1$, from ① we have

$$c_n > \frac{(n-1)^2}{n^2}c_{n-1} > \left(1 - \frac{2}{n}\right)c_{n-1} \geq c_{n-1} - \frac{2M}{n} > c_{n-1} - 1.$$

Therefore, $c_n \geq c_{n-1}$, which means $\{c_n\}$ is monotone non-decreasing after t_1 terms.

Note that $\{c_n\}$ has an upper bound M, so $c_n - c_{n-1}(n > t_1)$ has only a finite values of positive integers. Then there exists an integer $t \geq t_1$, such that $\{c_n\}$ will not increase after t terms, i.e., it becomes a constant sequence after the tth term.

Let $c_t = c_{t+1} = \cdots = K$. Then for any integer $n > t$, we have

$$a_n = (a_1 + \cdots + a_n) - (a_1 + \cdots + a_{n-1}) = n^2 K - (n-1)^2 K = (2n-1)K.$$

Therefore $K = \dfrac{a_n}{2n-1}$, and $b_n = a_{n+1} - a_n = (2n+1)K - (2n-1)K = 2K$, a constant.

The proof is completed. $\qquad\square$

11th Grade

First Day

(8:00 – 12:00; July 30, 2016)

1 As shown in Fig. 1.1, PAB and PCD are two secant lines of circle O, AD and BC intersect at point Q, T is a point on line BQ, line segment PT and circle O intersect at point K, and line QK and line PA intersect at point S.

Prove: If $ST // PQ$, then B, S, K, and T are concyclic.

Solution As shown in Fig. 1.2, let segment PQ and its extension intersect with circle O at E and F, respectively. From

$$\angle PQA > \angle ADC = \angle ABC$$

we know there exists point G on PQ, such that $\angle ABG = \angle PQA$. Then A, B, G, Q are four concyclic points. Therefore,

$$PQ \cdot PG = PA \cdot PB = PE \cdot PF,$$

and

$$\angle PGB = 180° - \angle BAD = 180° - \angle BCD = \angle PCB.$$

Then B, G, C, P are also four concyclic points, and

$$PQ \cdot QG = QC \cdot QB = QE \cdot QF.$$

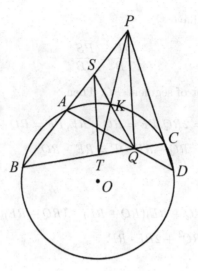

Fig. 1.1

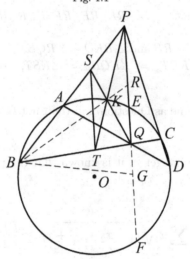

Fig. 1.2

Therefore,

$$PQ^2 = PQ \cdot PG - PQ \cdot QG = PE \cdot PF - QE \cdot QF.$$

Extend BK to intersect with PQ at point R. In $\triangle PBQ$, from Ceva's theorem we have

$$\frac{PR}{RQ} \cdot \frac{QT}{TB} \cdot \frac{BS}{SP} = 1. \qquad \qquad \text{①}$$

Since $ST // PQ$, we have

$$\frac{QT}{TB} = \frac{PS}{SB}. \qquad \textcircled{2}$$

So R is the midpoint of segment PQ. Then

$$PQ = 2RQ, PE = RQ + RE, PF = RQ + RF,$$
$$QE = RQ - RE, QF = RF - RQ,$$

Furthermore,

$$4RQ^2 = (RQ + RE)(RQ + RF) - (RQ - RE)(RF - RQ)$$
$$= 2RQ^2 + 2RE \cdot RF,$$

and that is

$$RQ^2 = RE \cdot RF \cdot RE \cdot RF = RK \cdot R.$$

Therefore, $RQ^2 = RK \cdot RB$ and $\triangle RKQ \backsim \triangle RQB$.

Consequently, $\angle KBT = \angle KQR = \angle KST$, so B, S, K, T are concyclic. $\qquad \square$

Remark   It is the same as Question 2 in the First Grade Grade Competition.

②   Given positive integer n, it is known that the product of n positive numbers $x_1, x_2, \ldots, x_n$ is 1.
Prove:

$$\sum_{i=1}^{n} x_i \sqrt{x_1^2 + x_2^2 + \cdots + x_i^2} \geq \frac{n+1}{2} \sqrt{n}.$$

Solution   From the given condition we have

$$\sum_{i=1}^{n} x_i \sqrt{x_1^2 + x_2^2 + \cdots + x_i^2}$$

$$\geq \sum_{i=1}^{n} x_i \cdot \frac{x_1 + x_2 + \cdots + x_i}{\sqrt{i}} = \sum_{i=1}^{n} \sum_{j=1}^{i} \frac{x_i x_j}{\sqrt{i}}$$

$$\geq \frac{1}{\sqrt{n}} \sum_{i=1}^{n} \sum_{j=1}^{i} x_i x_j = \frac{1}{2\sqrt{n}} \sum_{1 \leq j \leq i \leq n} 2 x_i x_j$$

$$= \frac{1}{2\sqrt{n}} \left(\sum_{i=1}^{n} x_i^2 + \left(\sum_{k=1}^{n} x_k \right)^2 \right)$$

$$\geq \frac{1}{2\sqrt{n}} [n \cdot \sqrt[n]{x_1^2 x_2^2 \ldots x_n^2} + (n \cdot \sqrt[n]{x_1 x_2 \ldots x_n})^2]$$

$$= \frac{n + n^2}{2\sqrt{n}} = \frac{n+1}{2} \sqrt{n}.$$

3 Given 4 points on the plane, they can be used as vertexes to form 4 triangles and as endpoints to formed 6 segments. If the areas of the 4 triangles are 4 different positive integers, and the lengths of the 6 segments are 6 different positive integers, we say the convex hull of these four points is a "lotus diagram".

(1) Give a specific "lotus diagram", including the areas of the corresponding 4 triangles and the lengths of corresponding 6 segments.

(2) Prove that there are infinitely many "lotus diagrams" on the plane, which are not similar to each other.

Note For a set S consisting of a finite number of points that are not collinear on the plane, its convex hull is a convex polygon area Ω (including the boundary and the interior): each vertex of Ω is a point in S, and every point in S belongs to Ω.

Solution (1) The two complete quadrilaterals shown in Fig. 3.1 and Fig. 3.2 are formed by concatenating four right-angled triangles whose sides are integers. The six sides of the obtained quadrilateral are positive integers different from each other, and the values of the areas are also mutually different positive integers, so these two quadrilaterals are lotus diagrams.

(2) Take any two Pythagorean triples (a, b, c) and (x, y, z), with

$$a^2 + b^2 = c^2, \quad x^2 + y^2 = z^2.$$

Make convex quadrilaterals $ABCD$ diagonally perpendicular to each other at point O, such that

$$OA = ay, \quad OB = ax, \quad OC = bx, \quad OD = by.$$

From the Pythagorean theorem we get that $AB = az$, $BC = cx$, $CD = bz$, $DA = cy$, so the distance between any two points in $A, B, C,$ and D is a

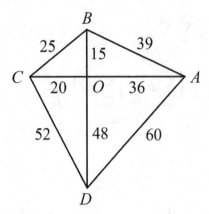

Fig. 3.1

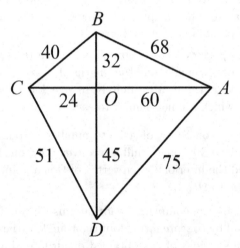

Fig. 3.2

positive integer. It is easy to know that there must be an even number in a and b, and an even number in x and y, so the areas of the four right-angled triangles $\triangle AOB$, $\triangle BOC$, $\triangle COD$, $\triangle DOA$ are all integers; then the areas of $\triangle ABC$, $\triangle BCD$, $\triangle CDA$ and $\triangle DAB$ must also be positive integers. Denote such convex quadrilateral $ABCD$ as $Q(a, b, c, x, y, z)$.

For any integer $t > 1$, note that $(2t, t^2 - 1, t^2 + 1)$ is a Pythagorean triple. Let

$$Q_t = Q(3, 4, 5, 2t, t^2 - 1, t^2 + 1) = A_t B_t C_t D_t.$$

We prove that integer $t > 1$ can be assigned infinitely many values, to make the convex hull of Q_t a lotus diagram. Let

$$\begin{cases} f_1(t) = A_t B_t = 3t^2 + 3, \\ f_2(t) = B_t C_t = 10t, \\ f_3(t) = C_t D_t = 4t^2 + 4, \\ f_4(t) = D_t A_t = 5t^2 - 5, \\ f_5(t) = A_t C_t = 3t^2 + 8t - 3, \\ f_6(t) = B_t D_t = 4t^2 + 6t - 4, \end{cases}$$

$$\begin{cases} g_1(t) = S_{\triangle A_t B_t C_t} = 3t(3t^2 + 8t - 3), \\ g_2(t) = S_{\triangle B_t C_t D_t} = 4t(4t^2 + 6t - 4), \\ g_3(t) = S_{\triangle C_t D_t A_t} = (2t^2 - 2)(3t^2 + 8t - 3), \\ g_4(t) = S_{\triangle D_t A_t B_t} = (3t^2 - 3)(2t^2 + 3t - 2). \end{cases}$$

For any (i, j) $(1 \le i < j \le 6)$, equation $f_i(t) = f_j(t)$ has at most two solutions in $\mathbb{N}^+$; and for any (i, j) $(1 \le i < j \le 4)$, equation $g_i(t) = g_j(t)$ also has at most two solutions in $\mathbb{N}^+$. Then there exists an infinite set $T \subseteq \mathbb{N}^+ \backslash \{1\}$, such that for every $t \in T$, $f_k(t)(k = 1, 2, \ldots, 6)$ are different from each other, and $g_l(t)(l = 1, 2, 3, 4)$ are also different from each other. At this time, $Q_t = A_t B_t C_t D_t$ becomes a lotus diagram.

Furthermore, when $t > 3$, it is easy to find that the ratio of the longest side to the shortest side of Q_t is

$$\varphi(t) = \frac{D_t A_t}{B_t C_t} = \frac{5t^2 - 5}{10t} = \frac{1}{2}\left(t - \frac{1}{t}\right),$$

From the monotonicity of $\varphi(t)$, we know that for any $t_1, t_2 \in T$, $3 < t_1 < t_2$, $\varphi(t_1) < \varphi(t_2)$. So Q_{t_1} is not similar to Q_{t_2}. Therefore, there are infinitely many of lotus diagrams that are not similar to each other.

The proof is completed. □

Remark   It is the same as Question 4 in the First Grade Grade Competition.

④   A substitute teacher led a group of students to go on an outing. The head teacher had told the teacher that two of students were mischievous at times (but which two students, the substitute teacher was not sure), and the other students always told the truth. Lost their way in the forest, and finally they gathered at a crossroad. The substitute teacher knew that their camp was on one of the four roads and was 20 minutes away from the junction, but they were not sure which road they were on. They must rush back to camp before dark.

(1) If the number of students in this group is 8, and there are 60 minutes before dark, please provide a plan to ensure that all students return to the camp before dark, and explain the reason.

(2) If the number of students in this group is 4, and there are still 100 minutes before dark, then is there a plan to ensure that all students return to the camp before dark? Prove your conclusion.

Solution   We use A, B, C, and D to indicate the four paths. In the first 40 minutes, the teacher should explored one of the four paths, say Path A, himself, and at the same time send students to explore the other three paths separately. If Path A was found by the teacher leading to the camp, the problem was successfully solved. Otherwise, the camp must be among the three paths B, C, and D.

(1) In the first 20 minutes, the teacher sent 3 students each on Paths B and C, and 2 students on Path D. In the next 20 minutes, everyone returned to the crossroads. For all possible reports, the teacher can press the table below and make the corresponding judgment (the teacher trusts a subgroup, which means trusting the judgment of the majority in this subgroup):

Subgroup B	Subgroup C	Subgroup D	Trusted subgroups
3:0	3:0	2:0	B, C
3:0	3:0	1:1	B, C
3:0	2:1	2:0	B, D
3:0	2:1	1:1	B, C
2:1	3:0	2:0	C, D
2:1	3:0	1:1	B, C
2:1	2:1	2:0	B, C, D
2:1	2:1	1:1	impossible

The last situation is impossible because if this happens, then there are at least three naughty students in the group, which is in conflict with the given conditions. With this table, the teacher then always knows the exact information of two paths in B, C, and D, which is enough for him to take students to go on the correct path and to reach the camp in the last 20 minutes.

(2) The conclusion is yes. Since there are 80 minutes for them to explore the route and return to the crossroad, one option is as follows:

In the first round, the teacher sent 4 students to explore Path B. If the report is in the form of 4: 0 or 3: 1, the judgment of most people can be trusted. Even if the answer is negative, the teacher can determine whether the camp is on the C route by himself in the second round.

If the report is in the 2: 2 form, then in the second round, the teacher sends a student S who answers "No" to explore Path C, and at the same time, he rechecks the B route himself. If he finds the camp, it succeeds; if the camp is not found, it means the 2 students who answered "Yes" in the first round are naughty, so S is not a naughty student, and the information of Path C fed back by S is credible.

So after two rounds of exploration, the teacher has either been able to determine the location of the camp, or can clearly deny the three paths A, B, and C, so that they can reach the camp from the crossroad in the last 20 minutes. □

Second Day
(8:00–12:00; July 31, 2016)

5 Given constant a satisfying $0 < \alpha \le 1$, prove:

(1) There exists constant $C(\alpha) > 0$, so that for all $x \ge 0$, $\ln(1+x) \le C(\alpha)x^\alpha$;

(2) for any two non-zero complex numbers z_1 and z_2,

$$\left| \ln \left| \frac{z_1}{z_2} \right| \right| \le C(\alpha) \left| \frac{z_1 - z_2}{z_2} \right|^\alpha + \left| \frac{z_2 - z_1}{z_1} \right|^\alpha,$$

where $C(\alpha)$ is the contant given in (1).

Solution　(1) Let $f(x) = C(\alpha)x^\alpha - \ln(1 + x)$, then for $x > 0$,

$$f'(x) = \alpha C(\alpha)x^{\alpha-1} - \frac{1}{1+x} = \frac{\alpha C(\alpha)x^\alpha(1+x) - x}{x(1+x)}. \qquad \text{(1)}$$

Note that $0 < \alpha \le 1 \le 1 + \alpha$, we have

$$x^\alpha(1+x) = x^\alpha + x^{1+\alpha} \ge \max\{x^\alpha, x^{1+\alpha}\} \ge x.$$

Let $C(\alpha)$ satisfy $\alpha C(\alpha) = 1$, and then from ① we have $f'(x) \ge 0$. Therefore,

$$f(x) \ge f(0) = 0 (x \ge 0).$$

So let $C(\alpha) = \dfrac{1}{\alpha}$, for $x \geq 0$, we always have

$$\ln(1 + x) \leq C(\alpha)x^{\alpha}.$$

(2) By the conclusion in (1), we have

$$\ln \left| \frac{z_1}{z_2} \right| = \ln \left| 1 + \frac{z_1 - z_2}{z_2} \right| \leq \ln \left(1 + \left| \frac{z_1 - z_2}{z_2} \right| \right) \leq C(\alpha) \cdot \left| \frac{z_1 - z_2}{z_2} \right|^{\alpha}.$$

In the same way,

$$- \ln \left| \frac{z_1}{z_2} \right| = \ln \left| \frac{z_2}{z_1} \right| \leq C(\alpha) \cdot \left| \frac{z_2 - z_1}{z_1} \right|^{\alpha}.$$

Therefore,

$$\left| \ln \left| \frac{z_1}{z_2} \right| \right| = \max \left\{ \ln \left| \frac{z_1}{z_2} \right|, - \ln \left| \frac{z_1}{z_2} \right| \right\} \leq C(\alpha) \left(\left| \frac{z_1 - z_2}{z_2} \right|^{\alpha} + \left| \frac{z_2 - z_1}{z_1} \right|^{\alpha} \right).$$

$\square$

6   As shown in Fig. 6.1, point I is the incenter of $\triangle ABC$, whose sides BC, CA, and AB is tangent to the inscribed circle at points D, E, F, respectively; lines BI, CI, DI intersect with EF at points M, N, K, respectively; lines BN and CM intersect at point P, line AK and BC intersect at point G; the line through I perpendicular to PG and the line through P perpendicular to PB intersect at point Q. Prove: line BI bisects line segment PQ.

Solution   As shown in Fig. 6.2, through K draw line parallel to BC intersecting with AB and AC at points R and S, respectively. Since $IK \perp RS$ and $IF \perp AB$, $\angle RKI = \angle RFI = 90°$, then I, K, S, E are four concyclic points. In the same way, I, K, S, E are also four concyclic points. Then

$$\angle IRK = \angle IFK = \angle IEK = \angle ISK.$$

Therefore, $KR = KS$. As $RS // BC$, G is the midpoint of BC. Since

$$\angle CNE = \angle AEF - \angle ECN = 90° - \frac{\angle BAC}{2} - \frac{\angle ACB}{2}$$

$$= \frac{\angle ABC}{2} = \angle FBI,$$

B, I, N, F are then four concyclic points. It is easy to see that B, D, I, F are also four concyclic points. Therefore, B, D, I, N, F are five concyclic points. Then $IN \perp BN$. In the same way, $IM \perp CM$. So I is the orthocenter of $\triangle PBC$, and $PI \perp BC$, i.e., P, I, D are collinear.

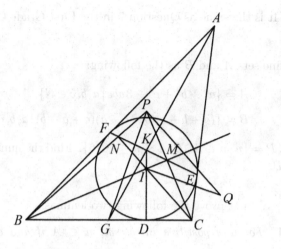

Fig. 6.1

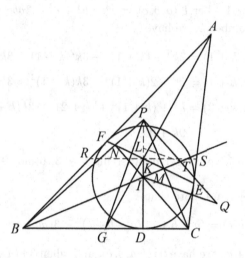

Fig. 6.2

Suppose BI and PQ intersect at T. From $\angle PIM = \angle PCB$, we know D, C, M, I are four concyclic points. Moreover,

$$\angle TPI = 90° - \angle BPD = \angle PBC.$$

Therefore, $\triangle TPI \backsim \triangle PBC$. Let L be the midpoint of PI. Then $\triangle TPL \backsim \triangle PBG$, and

$$\angle PTL + \angle GPT = \angle BPG + \angle GPT = \angle BPQ = 90°.$$

Therefore, $TL // QI$, and then $PT = TQ$, which means BI bisects PQ. $\square$

Remark It is the same as Question 7 in the First Grade Grade Competition.

7 Define sets A and B as the following:

$$A = \{a^3 + b^3 + c^3 - 3abc \,|\, a, b, c \in \mathbb{N}\},$$
$$B = \{(a + b - c)(b + c - a)(c + a - b) \,|\, a, b, c \in \mathbb{N}\}.$$

Let $P = \{n \,|\, n \in A \cap B,\ 1 \leq n \leq 2016\}$. Find the number of elements in P.

Solution We first prove the following two lemmas.

Lemma 1 *For any postitive integer n, $n \in A$ if and only if $n \not\equiv 3,$ $6 (\mathrm{mod}\, 9)$.*

Proof of lemma 1 Let $U(a, b, c) = a^3 + b^3 + c^3 - 3abc (a, b, c \in \mathbb{N})$. For every natural number k, we have

$$U(k, k, k+1) = 2k^3 + (k+1)^3 - 3k^2(k+1) = 3k + 1 \in A,$$
$$U(k, k+1, k+1) = k^3 + 2(k+1)^3 - 3k(k+1)^2 = 3k + 2 \in A,$$
$$U(k, k+1, k+2) = k^3 + (k+1)^3 + (k+2)^3 - 3k(k+1)(k+2)$$
$$= 9k + 9 \in A.$$

Then for any positive integer n satisfying $n \not\equiv 3,\ 6 (\mathrm{mod}\, 9)$, $n \in A$ always holds.

On the other hand, note that

$$U(a, b, c) = (a + b + c)[(a + b + c)^2 - 3(ab + bc + ca)].$$

When $3 \,|\, (a + b + c)$, we have $9 \,|\, U(a, b, c)$; and when $3 \nmid (a + b + c)$, we have $3 \nmid U(a, b, c)$. Therefore, there are not natural numbers a, b, c, such that $U(a, b, c) \equiv 3,\ 6 \,(\mathrm{mod}\, 9)$. In other words, for any $n \in A$, we always have $n \not\equiv 3,\ 6 \,(\mathrm{mod}\, 9)$.

Lemma 1 is then proved.

Lemma 2 *For any positive integer n, $n \in B$ if and only if $n \equiv 0, 1, 3, 5, 7 \,(\mathrm{mod}\, 8)$.*

Proof of lemma 2 Let

$$V(a, b, c) = (a + b - c)(b + c - a)(c + a - b)(a, b, c \in \mathbb{N}).$$

For any natural number k, we have

$$V(k+1, k+1, 1) = (2k+1) \cdot 1 \cdot 1 = 2k+1 \in B,$$

$$V(k+1, k+1, 2) = 2k \cdot 2 \cdot 2 = 8k \in B.$$

Then for any positive integer n satisfying $n \equiv 0, 1, 3, 5, 7 \pmod 8$, $n \in B$ always holds.

On the other hand, note that $a + b - c$, $b + c - a$, $c + a - b$ have the same parity. When they are even, we have $8 \mid V(a, b, c)$; and when they are odd, we have $2 \nmid V(a, b, c)$. Therefore, there are not natural numbers a, b, c, such that $V(a, b, c) \equiv 2, 4, 6 \pmod 8$. In other words, for any $n \in B$, we always have $n \not\equiv 2, 4, 6 \pmod 8$.

Lemma 2 is then proved.

Now we are going to find the number of elements in P.

According to lemmas 1 and 2, we have

$$P = \{n \mid 1 \leq n \leq 2016, \, n \equiv 0, 1, 3, 5, 7 \pmod 8, n \not\equiv 3, 6 \pmod 9\}.$$

We first find all the numbers of $\{1, 2, \ldots, 72\}$ that equal 0, 1, 3, 5, or 7 (mod 8):

$$1, \ 9, \ 17, 25, 33, 41, 49, 57, 65,$$

$$3, 11, 19, 27, 35, 43, 51, 59, 67,$$

$$5, 13, 21, 29, 37, 45, 53, 61, 69,$$

$$7, 15, 23, 31, 39, 47, 55, 63, 71,$$

$$8, 16, 24, 32, 40, 48, 56, 64, 72.$$

In the above five-row numbers, each row constitutes a completely residual system of modulo 9, so there are exactly $5 \times 7 = 35$ numbers that do not equal 3 or 6 (mod 9), and then they belong to P.

Since every $n \in P$ can expressed uniquely as $n = 72l + m(l \in \{0, 1, \ldots, 27\}$, $m \in \{1, 2, \ldots, 72\}$, and $n \in P$ if and only if $m \in P$. Therefore, the total number of elements in P is $35 \times 28 = 980$. $\qquad \Box$

8 Suppose positive integer sequence $\{a_n\}$ satisfies: $n^2 \mid (a_1 + a_2 + \cdots + a_n)$ and $a_n \leq (n + 2016)^2$, for any integer $n \geq 2016$. Let $b_n = a_{n+1} - a_n$. Prove: $\{b_n\}$ is a constant sequence after certain terms.

Solution Let $c_n = \dfrac{a_1 + a_2 + \cdots + a_n}{n^2}$ $(n = 1, 2, \ldots)$. Then by the given condtion, for any $n \geq 2016$, c_n is an integer, and

$$c_n = \frac{a_1 + \cdots + a_{n-1}}{n^2} + \frac{a_n}{n^2} = \frac{(n-1)^2}{n^2} c_{n-1} + \frac{a_n}{n^2}. \qquad \textcircled{1}$$

We will prove that $\{c_n\}$ is a constant sequence after certain terms.

First of all, letting $M = \max\{c_1, c_2, \ldots, c_{2016}, 10\,000\}$, we will prove by mathematical induction that M is an upper bound of $\{c_n\}$. When $k = 2016$, certainly $c_k \leq M$. Assume $c_{k-1} \leq M$ for $k > 2016$. Then by ① we have

$$c_k = \left(1 - \frac{2k-1}{k^2}\right) c_{k-1} + \frac{a_k}{k^2} \leq \left(1 - \frac{2k-1}{k^2}\right) M + \frac{(k+2016)^2}{k^2}$$

$$= M + 1 - \frac{1}{k^2}[(2k-1)M - 2016(2k+2016)] < M + 1. \qquad ②$$

The last inequality in ② holds because

$$(2k-1)M - 2016(2k+2016) > k \cdot 10000 - 2016 \cdot 3k > 0.$$

Since c_k is an integer, $c_k \leq M$. Therefore, M is an upper bound of $\{c_n\}$.

When integer $n > 2M = t_1$, from ① we have

$$c_n > \frac{(n-1)^2}{n^2} c_{n-1} > \left(1 - \frac{2}{n}\right) c_{n-1} \geq c_{n-1} - \frac{2M}{n} > c_{n-1} - 1.$$

Therefore, $c_n \geq c_{n-1}$, which means $\{c_n\}$ is monotone non-decreasing after t_1 terms.

Note that $\{c_n\}$ has an upper bound M, so $c_n - c_{n-1}(n > t_1)$ has only a finite values of positive integers. Then there exists an integer $t \geq t_1$, such that $\{c_n\}$ will not increase after t terms, i.e., it becomes a constant sequence after the tth term.

Let $c_t = c_{t+1} = \cdots = K$. Then for any integer $n > t$, we have

$$a_n = (a_1 + \cdots + a_n) - (a_1 + \cdots + a_{n-1}) = n^2 K - (n-1)^2 K = (2n-1)K.$$

Therefore $K = \dfrac{a_n}{2n-1}$, and $b_n = a_{n+1} - a_n = (2n+1)K - (2n-1)K = 2K$, a constant.

The proof is completed. □

Remark It is the same as Question 8 in the First Grade Competition.

China Southeastern Mathematical Olympiad

From July 28 to August 2, 2017, the 14th China Southeastern Mathematical Olympiad (CSMO) was held in the Yushan No.1 Middle School in Yushan County, Jiangxi Province, China. Nearly 12,000 players participated in the competition.

10th Grade
First Day
(8:00 – 12:00; July 30, 2017)

1 Let $x_i \in \{0, 1\}$ $(i = 1, 2, \ldots, n)$. If the value of function $f = f(x_1, x_2, \ldots, x_n)$ only takes 0 or 1, it is said that f is an n-ary Boolean function, and define $D_n(f) = \{(x_1, x_2, \ldots, x_n) \mid f(x_1, x_2, \ldots, x_n) = 0\}$.

(1) Find the number of n-ary Boolean functions.

(2) Let g be a 10-ary Boolean function, satisfying

$$g(x_1, x_2, \ldots, x_{10}) \equiv 1 + x_1 + x_1 x_2 + x_1 x_2 x_3$$
$$+ \cdots + x_1 x_2 \ldots x_{10} \pmod 2.$$

Find the number of elements in $D_{10}(g)$ and the value of
$$\sum_{(x_1, \ldots, x_{10}) \in D_{10}(g)} (x_1 + x_2 + \cdots + x_{10}).$$

Solution (1) There are totally 2^n value combinations that a group of x_1, $x_2, \ldots, x_n$ can take, and an n-ary Boolean function may take either 1 or 0 on each of them. Hence the number of n-ary Boolean functions is 2^{2^n}.

(2) Define $|D_{10}(g)|$ as the number of elements in $D_{10}(g)$, and * be either 1 or 0. Then we have

(3) For $(x_1, x_2, \ldots, x_{10}) = (1, 0, *, *, *, *, *, *, *, *)$, $g(x_1, x_2, \ldots, x_{10}) = 0$, and the number is $2^8 = 256$;

(4) for $(x_1, x_2, \ldots, x_{10}) = (1, 1, 1, 0, *, *, *, *, *, *)$, $g(x_1, x_2, \ldots, x_{10}) = 0$, and the number is $2^6 = 64$;

(5) for $(x_1, x_2, \ldots, x_{10}) = (1, 1, 1, 1, 1, 0, *, *, *, *)$, $g(x_1, x_2, \ldots, x_{10}) = 0$, and the number is $2^4 = 16$;

(6) for $(x_1, x_2, \ldots, x_{10}) = (1, 1, 1, 1, 1, 1, 1, 0, *, *)$, $g(x_1, x_2, \ldots, x_{10}) = 0$, and the number is $2^2 = 4$;

(7) for $(x_1, x_2, \ldots, x_{10}) = (1, 1, 1, 1, 1, 1, 1, 1, 1, 0)$, $g(x_1, x_2, \ldots, x_{10}) = 0$, and the number is 1.

Hence the number of elements in $D_{10}(g)$ is $256 + 64 + 16 + 4 + 1 = 341$, and

$$\sum_{(x_1, x_2, \ldots, x_{10}) \in D_{10}(g)} (x_1 + x_2 + \cdots + x_{10})$$

$$= 1 \times 256 + 128 \times 8 + 3 \times 64 + 32 \times 6 + 5 \times 16$$

$$+ 8 \times 4 + 7 \times 4 + 2 \times 2 + 9 = 1817.$$

2 As shown in Fig. 2.1, in acute angle $\triangle ABC$, $AB \neq AC$, K is the midpoint of the midline AD, $DE \perp AB$ at E, $DF \perp AC$ at F, lines KE, KF intersect BC at M and N, respectively, and O_1, O_2 are the circumcenters of $\triangle DEM$, $\triangle DFN$, respectively. Prove: $O_1O_2 \parallel BC$.

Solution   As shown is Fig. 2.2, take AD as the diameter to make circle K, and pass point A to make line $AG \parallel BC$ and intersect circle K at G. Then $\angle GAE = 180° - \angle ABC$, $\angle GAF = \angle ACB$. Hence

$$DE \cdot GF = DB\sin \angle ABC \cdot AD\sin \angle GAF = DB \cdot AD\sin \angle ABC \cdot \sin \angle ACB,$$

$$DF \cdot GE = DC\sin \angle ACB \cdot AD\sin \angle GAE = DC \cdot AD\sin \angle ACB \cdot \sin \angle ABC.$$

Furthermore, $DB = DC$, therefore $DE \cdot GF = DF \cdot GE$.

Through E and F make the tangents EP and FP of circle K, respectively. Suppose EF and DG intersect at point R, and lines ED, FD intersect

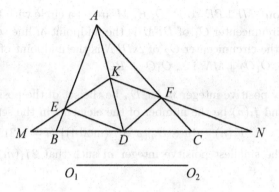

Fig. 2.1

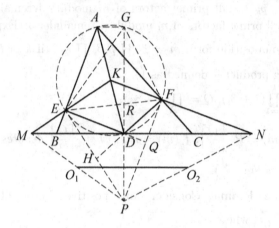

Fig. 2.2

FP, EP at points Q, H, respectively. Then in $\triangle DEF$,

$$\frac{ER}{RF} \cdot \frac{FH}{HD} \cdot \frac{DQ}{QE} = \frac{ED\sin\angle EDG}{FD\sin\angle FDG} \cdot \frac{EF\sin\angle FEH}{ED\sin\angle DEH} \cdot \frac{FD\sin\angle QFD}{EF\sin\angle QFE}$$

$$= \frac{\sin\angle EDG\sin\angle FED}{\sin\angle FDG\sin\angle DFE} = \frac{GE \cdot DF}{GF \cdot DE} = 1.$$

(as $\angle FEH = \angle QFE, \angle QFD = \angle FED, \angle DEH = \angle DFE$)

From the inverse of Seva's theorem, we know that the three lines of EH, FQ, and RD have the same point, i.e., points G, D, and P are collinear. Since $DG \perp AG$, $BC \parallel AG$, then $PD \perp BC$.

Furthermore, $ME \perp PE$, so P, D, E, M are on a circle with the diameter PM, so the circumcenter O_1 of DEM is the midpoint of line segment PM. Similary, the circumcenter O_2 of $\triangle DFN$ is the midpoint of PN. Therefore, $O_1 O_2 \parallel MN$, i.e., $O_1 O_2 \parallel BC$.

3   For any positive integer n, let D_n be the set of the positive divisors of n, and $f_i(n)$ be the number of the elements in the set

$$F_i(n) = \{a \in D_n \mid a \equiv i \pmod 4\} \ (i = 1, 2).$$

Find the smallest poisitve integer m such that $2f_1(m) - f_2(m) = 2017$.

Solution   Assume that m satisfies $2f_1(m) - f_2(m) = 2017$. Obviously $f_2(m)$ is not zero, so $2 \mid m$.

Let $p_1, \ldots, p_k$ be all prime factors of m modulo 4 remainder 1, and $q_1, \ldots, q_l$ be all prime factors of m modulo 4 remainder 3. Express m as a standard decomposition form $m = 2^\alpha \prod_{i=1}^{k} p_i \beta_i \prod_{j=1}^{k} q_j^{\gamma_j}$ (if k or l is zero, the corresponding product is defined as 1).

Let $P = \prod_{i=1}^{k} (1 + \beta_i)$, $Q = \prod_{j=1}^{l} (1 + \gamma_j)$.

Lemma $f_1(m) = P \cdot \left\lceil \dfrac{Q}{2} \right\rceil$, $f_2(m) = PQ$ (where $\left\lceil \dfrac{Q}{2} \right\rceil$ denotes the smallest integer not less than $\dfrac{Q}{2}$).

Proof of the lemma   Consider any positive factor of m, $a = 2^{\alpha'} \prod_{i=1}^{k} p_i^{\beta_i'} \prod_{j=1}^{l} q_j^{\gamma_j'}$, where

$$\alpha' \in \{0, 1, \ldots, \alpha\}, \ \beta_i' \in \{0, 1, \ldots, \beta_i\} \ (1 \le i \le k),$$

$$\gamma_j' \in \{0, 1, \ldots, \gamma_j\} \ (1 \le j \le l).$$

First calculate $f_2(m)$. Since $a \equiv 2 \pmod 4$ if and only if $\alpha' = 1$, it is obtained by the principle of multiplication,

$$f_2(m) = |F_2(m)| = 1 \cdot \prod_{i=1}^{k} (1 + \beta_i) \cdot \prod_{j=1}^{l} (1 + \gamma_j) = PQ.$$

Then consider $f_1(m)$. Note that $a \equiv 1 \pmod 4$ if and only if $\alpha' = 0$, $\sum_{j=1}^{l} \gamma_j' \equiv 0 \pmod 2$. We perform mathematical induction on l, to prove that the number of qualified $(\gamma_1', \gamma_2', \ldots, \gamma_l')$ is $\left\lceil \dfrac{Q}{2} \right\rceil$.

When $l = 0$, it is obvious that $Q = 1$, and then $\left\lceil \frac{Q}{2} \right\rceil = 1$. The conclusion is true.

Assuming that the conclusion is true for $l - 1$, consider the situation of l. There are the following two cases:

(a) If there is a $\gamma_j (1 \le j \le l)$ that is an odd number, we may assume it is γ_l. Then there are $\sum_{j=1}^{l-1} (1 + \gamma_j)$ ways to set values for $\gamma_1', \ldots, \gamma_{l-1}'$, and for each way there are exactly $\frac{1 + \gamma_l}{2}$ ways to select a $\gamma_l' \in \{0, 1, \ldots, \gamma_l\}$ such that $\gamma_1' + \cdots + \gamma_{l-1}' + \gamma_l' \equiv 0 \pmod 2$. Hence the number of the ways that meet the condition is

$$\frac{1 + \gamma_l}{2} \times \sum_{j=1}^{l-1} (1 + \gamma_j) = \frac{Q}{2} = \left\lceil \frac{Q}{2} \right\rceil.$$

(b) If $\gamma_j (1 \le j \le l)$ are all even numbers, then according to the inductive hypothesis, there are $\lambda_1 = \left\lceil \frac{1}{2} \prod_{j=1}^{l-1} (1 + \gamma_j) \right\rceil$ ways to set values for $\gamma_1', \ldots, \gamma_{l-1}'$ with $\gamma_l' = 0$ such that $\gamma_1' + \cdots + \gamma_{l-1}' + \gamma_l' \equiv 0 \pmod 2$. On the other hand, we may first set any values to $\gamma_1', \ldots, \gamma_{l-1}'$, and then select a $\gamma_l' \in \{1, 2, \ldots, \gamma_l\}$ such that $\gamma_1' + \cdots + \gamma_{l-1}' + \gamma_l' \equiv 0 \pmod 2$; it is easy to see that the number of ways to do so is $\lambda_2 = \frac{\gamma_l}{2} \cdot \sum_{j=1}^{l-1} (1 + \gamma_j)$.

Hence, the total number is $\lambda_1 + \lambda_2 = \left\lceil \frac{Q}{2} \right\rceil$.

Then by mathematical induction, we know that, for any l, the number of qualified ways to set values for $(\gamma_1', \gamma_2', \ldots, \gamma_l')$ is $\left\lceil \frac{Q}{2} \right\rceil$.

Finally, by the principle of multiplication, we have $f_1(m) = |F_1(m)| = 1 \cdot \prod_{i=1}^{k} (1 + \beta_i) \cdot \left\lceil \frac{Q}{2} \right\rceil = P \cdot \left\lceil \frac{Q}{2} \right\rceil$. The lemma is then proved.

By the lemma, we have

$$2017 = 2f_1(m) - f_2(m) = P \cdot \left[2 \left\lceil \frac{Q}{2} \right\rceil - Q \right]. \qquad \textcircled{1}$$

Here Q must be an odd number, so $2 \left\lceil \frac{Q}{2} \right\rceil - Q = 1$. Hence $P = \prod_{i=1}^{k} (1 + \beta_i) = 2017$. Note that 2017 is a prime number. Then $k = 1, \beta_1 = 2016$. We have

$$m = 2^\alpha p_1^{2016} \prod_{j=1}^{l} q_j^{\gamma_j} \ge 2 \cdot 5^{2016}.$$

On the other hand, when $m = 2 \cdot 5^{2016}$, we have $P = 2017, Q = 1$, which satisfy ①. Therefore, the required minimum number is $m = 2 \cdot 5^{2016}$. □

4 Suppose real numbers $a_1, a_2, \ldots, a_{2017}$ satisfy $a_1 = a_{2017}$, and

$$|a_i + a_{i+2} - 2a_{i+1}| \le 1, \quad i = 1, 2, \ldots, 2015.$$

Define $M = \max\limits_{1 \le i < j \le 2017} |a_i - a_j|$. Find the maximum value of M.

Solution We will prove that the maximum value of M is $\dfrac{1008^2}{2}$. Suppose

$$|a_{i_0} - a_{j_0}| = \max\limits_{1 \le i < j \le 2017} |a_i - a_j| = M, \quad 1 < i_0 < 2017.$$

Then a_{i_0} must be either the maximum or the minimum in $a_1, a_2, \ldots, a_{2017}$, so, with the given condition, we have

$$0 \le (a_{i_0} - a_{i_0-1})(a_{i_0} - a_{i_0+1}) \le \left(\frac{a_{i_0} - a_{i_0-1} + a_{i_0} - a_{i_0+1}}{2}\right)^2 \le \frac{1}{4}.$$

Hence,

$$\max\{|a_{i_0} - a_{i_0-1}|, |a_{i_0} - a_{i_0+1}|\} \le |a_{i_0-1} + a_{i_0+1} - 2a_{i_0}| \le 1,$$

$$\min\{|a_{i_0} - a_{i_0-1}|, |a_{i_0} - a_{i_0+1}|\} \le \frac{1}{2}.$$

Case 1: $j_0 = 1$ or 2017.

(i) For $i_0 = 1009$. If $|a_{i_0} - a_{i_0-1}| \le \dfrac{1}{2}$, we have

$$|a_{i_0-1} - a_{i_0-2}| \le |a_{i_0} + a_{i_0-2} - 2a_{i_0-1}| + |a_{i_0} - a_{i_0-1}| \le 1 + \frac{1}{2},$$

$$|a_{i_0-2} - a_{i_0-3}| \le |a_{i_0-1} + a_{i_0-3} - 2a_{i_0-2}| + |a_{i_0-1} - a_{i_0-2}|$$

$$\le 1 + 1 + \frac{1}{2} = 2 + \frac{1}{2},$$

$\ldots \ldots$

$$|a_2 - a_1| \le |a_3 + a_1 - 2a_2| + |a_3 - a_2| \le 1 + 1006 + \frac{1}{2} = 1007 + \frac{1}{2}.$$

Therefore,

$$|a_{i_0} - a_1| \le |a_{i_0} - a_{i_0-1}| + |a_{i_0-1} - a_{i_0-2}| + \cdots + |a_2 - a_1|$$

$$\le \frac{1}{2} + \left(1 + \frac{1}{2}\right) + \cdots + \left(1007 + \frac{1}{2}\right) = \frac{1008^2}{2}.$$

If $|a_{i_0} - a_{i_0+1}| \leq \frac{1}{2}$, in the same way we have

$$|a_{i_0} - a_{2017}| \leq \frac{1008^2}{2}.$$

(ii) For $i_0 < 1009$, in the same way we have

$$|a_{i_0} - a_1| \leq 1 + 2 + \cdots + 1007 \leq \frac{1008^2}{2}.$$

(iii) For $i_0 \geq 1010$, we have $|a_{2017} - a_{i_0}| \leq 1 + 2 + \cdots + 1007 \leq \frac{1008^2}{2}.$

Case 2: $1 < j_0 < 2017$.

(i) For $j_0 - i_0 = 1008$. If $|a_{j_0} - a_{j_0-1}| \leq \frac{1}{2}$, then we have

$$|a_{j_0} - a_{i_0}| \leq |a_{i_0} - a_1| + |a_{2017} - a_{j_0}|$$

$$\leq \frac{(i_0 - 1)^2}{2} + \frac{(2017 - j_0)(2018 - j_0)}{2} \leq \frac{1008^2}{2}.$$

If $|a_{i_0} - a_{i_0+1}| \leq \frac{1}{2}$, then we have

$$|a_{j_0} - a_{i_0}| \leq |a_{i_0} - a_1| + |a_{2017} - a_{j_0}| \leq \frac{(i_0 - 1)i_0}{2} + \frac{(2017 - j_0)^2}{2} \leq \frac{1008^2}{2}.$$

(ii) For $|j_0 - i_0| < 1008$, we have

$$|a_{j_0} - a_{i_0}| \leq \frac{1007 \times 1008}{2} \leq \frac{1008^2}{2}.$$

(iii) For $|j_0 - i_0| > 1008$, we have

$$|a_{j_0} - a_{i_0}| \leq |a_{i_0} - a_1| + |a_{2017} - a_{j_0}| \leq \frac{(i_0 - 1)^2}{2} + \frac{(2017 - j_0)^2}{2}$$

$$\leq \frac{(2016 - j_0 + i_0)^2}{2} \leq \frac{1008^2}{2}.$$

In summary, we get $M \leq \frac{1008^2}{2}$.

On the other hand, let $a_n = \frac{(1009 - n)^2}{2}$, $n = 1, 2, \ldots, 2017$. We have

$$|a_{1009-a_1}| = \frac{1008^2}{2}.$$

Therefore, $M = \frac{1008^2}{2}$. $\qquad\qquad\qquad\square$

Second Day
(8:00 – 12:00; July 31, 2017)

5 As shown in Fig. 5.1, in the inscribed quadrilateral $ABCD$ of circle O, the diagonals AC and BD are perpendicular to each other. M and N are the midpoints of arc $\overset{\frown}{ADC}$ and $\overset{\frown}{ABC}$, respectively. The diameter passing point D intersects with chord AN at G. And K is a point on side CD, satisfying $GK \parallel NC$. Proof: $BM \perp AK$.

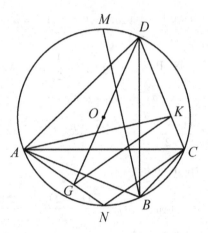

Fig. 5.1

Solution As shown in Fig. 5.2, connecting NB, since $GK \parallel NC$ and A, N, C, D are concyclic, we have

$$\angle AGK = \angle ANC = 180° - \angle ADC.$$

Then A, G, K, D are concyclic, and $\angle AKG = \angle ADG$. Furthermore, $DB \perp AC$. so $\angle BDC = 90° - \angle ACD = \angle ADG$. Therefore,

$$\angle BNC = \angle BDC = \angle ADG = \angle AKG.$$

Combing it with the fact that $GK \parallel NC$, we have $NB \parallel AK$. Furthermore, since M, N are antipodal points on circle O, then $BM \perp NB$. Hence, $BM \perp AK$. □

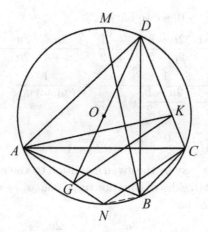

Fig. 5.2

6 It is known that real number sequence $\{a_n\}$ satisfies $a_1 = \dfrac{1}{2}$, $a_2 = \dfrac{3}{8}$, and

$$a_{n+1}^2 + 3a_n a_{n+2} = 2a_{n+1}(a_n + a_{n+2}) \quad (n = 1, 2, \ldots).$$

(1) Find the general term formula of $\{a_n\}$;

(2) prove $0 < a_n < \dfrac{1}{\sqrt{2n+1}}$.

Solution 1 (1) From $a_{n+1}^2 + 3a_n a_{n+2} = 2a_{n+1}(a_n + a_{n+2})$, we have

$a_n a_{n+2} - a_{n+1}^2 = 2a_{n+1}(a_n + a_{n+2}) - 2a_n a_{n+2} - 2a_{n+1}^2$, and that is

$$a_{n+1}(a_n - a_{n+1}) - a_n(a_{n+1} - a_{n+2}) = 2(a_n - a_{n+1})(a_{n+1} - a_{n+2}). \quad (1)$$

Obviously, $(a_n - a_{n+1})(a_{n+1} - a_{n+2}) \neq 0$. Then

$$\frac{a_{n+1}}{a_{n+1} - a_{n+2}} - \frac{a_n}{a_n - a_{n+1}} = 2.$$

Therefore,

$$\frac{a_n}{a_n - a_{n+1}} = \frac{a_1}{a_1 - a_2} + 2(n-1) = 2n + 2.$$

That is

$$a_{n+1} = \frac{2n+1}{2n+2} a_n.$$

Then we have

$$a_n = \frac{2n-1}{2n} a_{n-1} = \frac{2n-1}{2n} \cdot \frac{2n-3}{2n-2} \cdots \cdots \frac{3}{4} a_1 = \frac{(2n-1)!!}{(2n)!!}.$$

(2) Obviously $a_n > 0$, so we have

$$a_n = \frac{2n-1}{2n} \cdot \frac{2n-3}{2n-2} \cdots \frac{3}{4} \cdot \frac{1}{2} < \frac{2n}{2n+1} \cdot \frac{2n-2}{2n-1} \cdots \frac{4}{5} \cdot \frac{2}{3}$$

$$= \frac{1}{\dfrac{2n+1}{2n} \cdot \dfrac{2n-1}{2n-2} \cdots \dfrac{5}{4} \cdot \dfrac{3}{2}} = \frac{1}{(2n+1)a_n}.$$

Hence, $a_n^2 < \dfrac{1}{2n+1}$, i.e., $0 < a_n < \dfrac{1}{\sqrt{2n+1}}$.

Solution 2   (1) It is easy by the given condition to know that $a_n \neq 0$. Then we can divide the both sides of equation $a_{n+1}^2 + 3a_n a_{n+2} = 2a_{n+1}(a_n + a_{n+2})$ by $a_n a_{n+1}$ to obtain that

$$\frac{a_{n+1}}{a_n} + \frac{3a_{n+2}}{a_{n+1}} = 2 + \frac{2a_{n+2}}{a_n}.$$

Let $b_n = \dfrac{a_{n+1}}{a_n}$. Then we get $b_n + 3b_{n+1} = 2 + 2b_n b_{n+1}$, and that is

$$b_{n+1} - b_n = 2(b_n - 1)(b_{n+1} - 1). \qquad \qquad ②$$

Obviously, $b_n \neq 1$, so we get

$$\frac{1}{b_{n+1} - 1} - \frac{1}{b_n - 1} = -2.$$

Then we have

$$\frac{1}{b_n - 1} = \frac{1}{b_1 - 1} - 2(n-1) = \frac{1}{b_1 - 1} - 2(n-1) = -2n - 2.$$

Therefore, $a_{n+1} = \dfrac{2n+1}{2n+2} a_n$. Hence,

$$a_n = \frac{2n-1}{2n} a_{n-1} = \frac{2n-1}{2n} \cdot \frac{2n-3}{2n-2} \cdots \cdots \frac{3}{4} a_1 = \frac{(2n-1)!!}{(2n)!!}.$$

(3) Obviously, $a_n > 0$. When $n = 1$, we have $a_1 = \dfrac{1}{2} < \dfrac{1}{\sqrt{3}}$. Assume when $n = k$, we have $a_k < \dfrac{1}{\sqrt{2k+1}}$ Then

$$a_{k+1} = \frac{2k+1}{2k+2} a_k < \frac{2k+1}{2k+2} \cdot \frac{1}{\sqrt{2k+1}} = \frac{\sqrt{2k+1}}{2k+2} < \frac{1}{\sqrt{2k+3}}.$$

Therefore, by mathematical induction, we have

$$0 < a_n < \frac{1}{\sqrt{2n+1}}. \qquad \qquad \square$$

7 Let m be a positive integer. For $k = 1, 2, \ldots$, define $a_k = \dfrac{(2km)!}{3^{(k-1)m}}$. Prove: there are infinitely many integer terms and infinitely many non-integer terms in sequence $a_1, a_2, \ldots$.

Solution For any given positive integer t, take $r_t \in \{0, 1, 2, \ldots, m-1\}$ such that $r_t \equiv -3^t \pmod{m}$. Let $k_t = \dfrac{3^t + r_t}{m}$. We will prove that a_{k_t} is an integer, and $a_{k_t - 1}$ is a non-integer.

Let α_t, β_t denote the powers of the prime factor 3 in $(2k_t m)!$, $((2k_t - 1)m)!$, respectively. On the one hand, as $2k_t m = 2 \cdot (3^t + r_t) \geq 2 \cdot 3^t$, we have

$$\alpha_t = \sum_{i=1}^{\infty} \left[\frac{2k_t m}{3^i} \right] \geq \sum_{i=1}^{\infty} \left[\frac{2 \cdot 3^t}{3^i} \right]$$

$$= 2(3^{t-1} + 3^{t-2} + \cdots + 3 + 1) = 3^t - 1 \geq 3^t - (m - r_t) = m(k_t - 1).$$

Hence, $3^{(k_t - 1)m} \mid 2k_t m)!$, i.e., a_{k_t} is an integer.

On the other hand, since

$$2(k_t - 1)m = 2 \cdot (3^t + r_t - m) \leq 2 \cdot (3^t - 1),$$

we have

$$\beta_t = \sum_{i=1}^{\infty} \left[\frac{2(k_t - 1)m}{3^i} \right] \leq \sum_{i=1}^{\infty} \left[\frac{2 \cdot (3^t - 1)}{3^i} \right] = (2 \cdot 3^{t-i} - 1)$$

$$= 2(3^{t-1} + 3^{t-2} + \cdots + 3 + 1) - t = 3^t - 1 - t.$$

Hence, when $t \geq 2m$, $\beta_t < 3^t - 2m \leq k_t m - 2m$. Therefore, $3^{(k_t - 2)m}$ cannot divide $(2(k_t - 1)m)!$, so $a_{k_t - 1}$ is not an integer.

Since there are infinitely many t's meeting the requirement, the proposition is proved. □

8 Let $S = \{(a, b) \mid a \in \{1, 2, \ldots, m\}, b \in \{1, 2, \ldots, n\}\}$, with positive integers $m \geq 2$, $n \geq 3$, and A be the subset of S. Suppose there are not any x_1, x_2, y_1, y_2, y_3, satisfying $x_1 < x_2, y_1 < y_2 < y_3$, $(x_1, y_1), (x_1, y_2), (x_1, y_3), (x_2, y_2) \in A$. Find the maximum number of elements in A.

Solution 1 If there is a point (x_1, y_2) in a set X such that there are still points $(x_1, y_3), (x_1, y_1), (x_2, y_2)$ in X that are above, below, and to the right of (x_1, y_2), respectively, X is then called a "central point set". Hence, the problem is equivalent to determining how many points can be taken out of S so that the set A composed of these points is not a central point set.

Firstly, we take all the points with the ordinate 1 or n and the points with the abscissa m from S, to form set A_0. Obviously, A_0 is not a center point set, and we have $|A_0| = 2m + n - 2$.

Next, we will prove: if $A \subseteq S$, $|A| \geq 2m + n - 1$, then A is a central point set. For $1 \leq i \leq n$, assume there are k_i points with the ordinate i in A. If $k_i \geq 2$, we color the points red in row i except for the rightmost one. Then there are at least $(k_1 - 1) + \cdots + (k_n - 1) = |A| - n$ red points in A.

For $1 \leq j \leq m - 1$, assume there are l_j points with the abscissa j. If $l_j \geq 3$, we color the points blue in column j except for the top and bottom points, and also color all the points blue with the abscissa m. Then there are at least $(l_1 - 2) + \cdots + (l_{m-1} - 2) + l_m = |A| - 2m + 2$ blue points in A.

Note that $(|A| - n) + (|A| - 2m + 2) = (|A| - 2m - n + 1) + |A| + 1 > |A|$. Hence, there is a point in A being colored both red and blue: it is easy to check that there are other points in A which are above, below, and to the right of this point, respectively. Therefore, A is a central point set.

In summary, the maximum number of elements in A is $2m + n - 2$.

Solution 2 Consider S as an n-row m-column matrix S_{nm}, with $m \geq 1$, $n \geq 2$. We have learnt from Solution 1 that there is a subset A of S_{nm} that is not a central poinit set and $|A| = 2m + n - 2$. Hence, we only need to prove this proposition: if A is subset of S_{nm} and $|A| \geq 2m + n - 1$, then A is a central point set.

We perform mathematical induction on $n + m$. When $m = 1$ or $n = 2$, the proposition is obviously true. Assuming $m \geq 2$, $n \geq 3$, the proposition is true for $n + m - 1$. Now we consider $n + m$. If there is a row in S_{nm}, where only 1 point belonging to A exists, then droping this row, consider $S_{n-1,m}$. Since $|A \cap S_{n-1,m}| \geq |A| - 1 \geq 2m + (n - 1) - 1$, by induction we know $A \cap S_{n-1,m}$ is a central point set, and so is A.

If there are at least 2 points in A in each row of S_{nm}, consider the leftmost column of S_{nm}. If there are 3 points in A in this column, let they be u, v and w, from top to bottom. Since there must be $t \in A$ in the row of v and is to the right of v, so A is a center point set. If this column has at most 2 points in A, remove this column and consider S_{nm-1}. Since $|A \cap S_{nm-1}| \geq |A| - 2 \geq 2(m - 1) + n - 1$, by induction we know $A \cap S_{nm-1}$ is a central point set, and so is A. Therefore, by mathematical induction, the proposition is true.

In summary, the maximum number of elements in A is $2m + n - 2$. $\square$

11th Grade
First Day
(8:00 – 12:00; July 30, 2017)

1 As shown in Fig. 1.1, in acute angle $\triangle ABC$, $AB \neq AC$, K is the midpoint of the midline AD, $DE \perp AB$ at E, $DF \perp AC$ at F, lines KE, KF intersect BC at M and N, respectively, and O_1, O_2 are the circumcenters of $\triangle DEM$, $\triangle DFN$, respectively. Prove: $O_1O_2 // BC$.

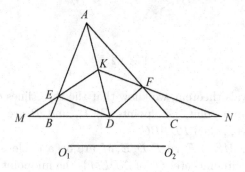

Fig. 1.1

Solution   As shown is Fig. 1.2, take AD as the diameter to make circle K, and pass point A to make line $AG // BC$ and intersect circle K at G. Then $\angle GAE = 180° - \angle ABC$, $\angle GAF = \angle ACB$. Hence

$$DE \cdot GF = DB \sin \angle ABC \cdot AD \sin \angle GAF = DB \cdot AD \sin \angle ABC \cdot \sin \angle ACB,$$

$$DF \cdot GE = DC \sin \angle ACB \cdot AD \sin \angle GAE = DC \cdot AD \sin \angle ACB \cdot \sin \angle ABC.$$

Furthermore, $DB = DC$, therefore $DE \cdot GF = DF \cdot GE$.

Through E and F make the tangents EP and FP of circle K, respectively. Suppose EF and DG intersect at point R, and lines ED, FD intersect FP, EP at points Q, H, respectively. Then in $\triangle DEF$,

$$\frac{ER}{RF} \cdot \frac{FH}{HD} \cdot \frac{DQ}{QE} = \frac{ED \sin \angle EDG}{FD \sin \angle FDG} \cdot \frac{EF \sin \angle FEH}{ED \sin \angle DEH} \cdot \frac{FD \sin \angle QFD}{EF \sin \angle QFE}$$

$$= \frac{\sin \angle EDG \sin \angle FED}{\sin \angle FDG \sin \angle DFE} = \frac{GE \cdot DF}{GF \cdot DE} = 1.$$

(as $\angle FEH = \angle QFE$, $\angle QFD = \angle FED$, $\angle DEH = \angle DFE$)

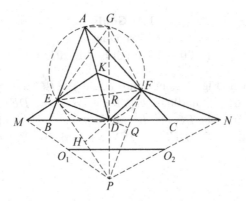

Fig. 1.2

From the Seva's theorem, we know that the three lines of EH, FQ, and RD have the same point, i.e., points G, D, and P are collinear. Since $DG \perp AG$, $BC \parallel AG$, then $PD \perp BC$.

Furthermore, $ME \perp PE$, so P, D, E, M are on a circle with PM as the diameter, so the circumcenter O_1 of $\triangle DEM$ is the midpoint of line segment PM.

Similarly, the circumcenter O_2 of $\triangle DFN$ is the midpoint of PN.

Therefore, $O_1 O_2 \parallel MN$, i.e., $O_1 O_2 \parallel BC$.   □

2 Let $x_i \in \{0, 1\}$ $(i = 1, 2, \ldots, n)$. If the value of function $f = f(x_1, x_2, \ldots, x_n)$ only takes 0 or 1, then it is said that f is an n-ary Boolean function, and define $D_n(f) = \{(x_1, x_2, \ldots, x_n) \mid f(x_1, x_2, \ldots, x_n) = 0\}$.

(1) Find the number of n-ary Boolean functions.

(2) Let g be a n-ary Boolean function, satisfying

$$g(x_1, x_2, \ldots, x_{10}) \equiv 1 + x_1 + x_1 x_2 + x_1 x_2 x_3$$

$$+ \cdots + x_1 x_2 \cdots x_n \pmod{2}.$$

Find the number of elements in $D_n(g)$, and the maximum n such that $\displaystyle\sum_{(x_1, \ldots, x_n) \in D_n(g)} (x_1 + x_2 + \cdots + x_n) 2017$.

Solution 1 (1) There are totally 2^n value combinations that a group of $x_1, x_2, \ldots, x_n$ can take, and an n-ary Boolean function may take either 1 or 0 on each of them. Hence the number of n-ary Boolean functions is 2^{2^n}.

(2) Define $|D_n(g)|$ as the number of elements in $D_n(g)$. Obviously $|D_1(g)| = 1$, $|D_2(g)| = 1$. Furthermore, from

$$g(x_1, x_2, \ldots, x_n) \equiv 1 + x_1 + x_1x_2 + x_1x_2x_3 + \cdots + x_1x_2 \cdots x_n \pmod 2$$

$$= (1 + x_1(1 + x_2 + x_2x_3 + \cdots + x_2x_3 \cdots x_n)) \pmod 2,$$

we have $|D_n(g)| = 2^{n-1} - |D_{n-1}(g)|$, $n = 3, 4, \ldots$, and then

$$|D_n(g)| = 2^{n-1} - 2^{n-2} + |D_{n-2}(g)| = \cdots = 2^{n-1} - 2^{n-2} + 2^{n-3}$$

$$+ \cdots + (-1)^{n-2}2 + (-1)^{n-1}.$$

Hence,

$$|D_n(g)| = \frac{2^n + (-1)^{n+1}}{3}, \quad n = 1, 2, \ldots.$$

The number of elements in $D_n(g)$ is then $\dfrac{2^n + (-1)^{n+1}}{3}$.

Let $c_n = \displaystyle\sum_{(x_1, x_2, \ldots, x_n) \in D_n(g)} (x_1 + x_2 + \cdots + x_n)$. It is obvious

$$g(x_1, x_2, \ldots, x_{n-1}, 0) \equiv 1 + x_1 + x_1x_2 + x_1x_2x_3$$

$$+ \cdots + x_1x_2 \cdots x_{n-1} \pmod 2, \text{ and}$$

$$g(x_1, x_2, \ldots, x_{n-1}, 1) \equiv 1 + x_1 + x_1x_2 + x_1x_2x_3$$

$$+ \cdots + x_1x_2 \cdots x_{n-1} + x_1x_2 \cdots x_{n-1} \pmod 2.$$

Then if $x_1x_2 \ldots x_{n-1} = 1$,

$$g(x_1, x_2, \ldots, x_{n-1}, 1) \equiv (n+1) \pmod 2 = \begin{cases} 0, & n \text{ is odd}, \\ 1, & n \text{ is even}. \end{cases}$$

If $x_1x_2 \ldots x_{n-1} = 0$, then

$$g(x_1, x_2, \ldots, x_{n-1}, 1) \equiv 1 + x_1 + x_1x_2 + x_1x_2x_3$$

$$+ \cdots + x_1x_2 \cdots x_{n-1} \pmod 2 = 0$$

is equivalent to

$$g(x_1, x_2, \ldots, x_{n-1}) \equiv 1 + x_1 + x_1x_2 + x_1x_2x_3 + \cdots$$

$$+ x_1x_2 \ldots x_{n-1} \pmod 2 = 0.$$

Therefore, when $n \geq 2$, we have

$$c_n = \begin{cases} n + 2c_{n-1} + |D_{n-1}(g)|, & n, \text{ is odd}, \\ 2c_{n-1} + |D_{n-1}(g)| - n, & n \text{ is even}. \end{cases}$$

Obviously, $c_1 = 1$, $c_2 = 1$. Then, when $n = 2m$, we have

$$c_{2m} = 2c_{2m-1} + \frac{2^{2m-1} + (-1)^{2m}}{3} - 2m = -\frac{1}{3} + \frac{4}{3} \cdot 2^{2m-2}$$
$$+ (2m - 2) + 4c_{2m-2} \tag{1}$$
$$= \frac{(3m + 1)4^{m} - (6m+1)}{9}, \quad m = 1, 2, 3, \ldots.$$

By the same reason, when $n = 2m + 1$, we have

$$c_{2m+1} = 2m + 1 + 2c_{2m} + \frac{2^{2m} + (-1)^{2m+1}}{3} = \frac{1}{3} + \frac{4}{3} \cdot 2^{2m-1}$$
$$- (2m - 1) + 4c_{2m-1} \tag{2}$$
$$= \frac{(6m + 5)4^{m} + 6m + 4}{9}, \quad m = 0, 1, 2, \ldots.$$

By ①, ②, we get

$$c_9 = 828 < 3986 = c_{11}, \quad c_{10} = 1817 < 8643 = c_{12}.$$

Therefore, the maximum positive integer n satisfying $\displaystyle\sum_{x_1, x_2, \ldots, x_n) \in D_n(g)} (x_1 + x_2 + \cdots + x_n) \leq 2017$ is $n = 10$.

Solution 2 (1) It is the same with what is in Solution 1.

(2) Define $|D_n(g)|$ as the number of elements in $D_n(g)$, and $*$ be either 1 or 0 in the following. When $n = 2m$, there are 2^{2m-2} elements in $D_n(g)$, satisfying

$$(x_1, x_2, \ldots, x_{2m}) = (1, 0, *, *, \ldots, *);$$

there are 2^{2m-4} elements in $D_n(g)$, satisfying

$$(x_1, x_2, \ldots, x_{2m}) = (1, 1, 1, 0, *, *, \ldots, *);$$

there are 2^{2m-6} elements in $D_n(g)$, satisfying

$$(x_1, x_2, \ldots, x_{2m}) = (1, 1, 1, 1, 1, 0, *, *, \ldots, *);$$

there are 2^2 elements in $D_n(g)$, satisfying

$$(x_1, x_2, \ldots, x_{2m}) = (1, 1, \ldots, 1, 0, *, *);$$

there is just one element in $D_n(g)$, satisfying

$$(x_1, x_2, \ldots, x_{2m}) = (1, 1, \ldots, 1, 0).$$

Hence, the number of elements in $D_n(g)$ is

$$|D_{2m}(g)| = 2^{2m-2} + 2^{2m-4} + \cdots + 2^2 + 1 = \frac{2^{2m} - 1}{3}.$$

In the same way, when $n = 2m + 1$, the number of elements in $D_{2m+1}(g)$ is

$$|D_{2m+1}(g)| = 2^{2m-1} + 2^{2m-3} + \cdots + 2^{1+1} = \frac{2^{2m+1} + 1}{3}.$$

In summary, $|D_n(g)| = \dfrac{2^n + (-1)^{n+1}}{3}$, $n = 1, 2, \ldots$.

Let $c_n = \displaystyle\sum_{(x_1, x_2, \ldots, x_n) \in D_n(g)} (x_1 + x_2 + \cdots + x_n)$.

In the case $n = 2m$, we have the following result.

For $(x_1, x_2, \ldots, x_{2m}) = (1, 0, *, *, \ldots, *)$, there are

$$1 \times 2^{2m-2} + 2^{2m-3} \times (2m - 2) = m \times 2^{2m-2}$$

number 1s in the solutions of $g(x_1, x_2, \ldots, x_{2m}) = 0$; for $(x_1, x_2, \ldots, x_{2m}) = (1, 1, 1, 0, *, *, \ldots, *)$, there are

$$3 \times 2^{2m-4} + 2^{2m-5} \times (2m - 4) = (m + 1) \times 2^{2m-4}$$

number 1s in the solutions of $g(x_1, x_2, \ldots, x_{2m}) = 0$; for $(x_1, x_2, \ldots, x_{2m}) = (1, 1, 1, 1, 1, 0, *, *, \ldots, *)$, there are

$$5 \times 2^{2m-6} + 2^{2m-7} \times (2m - 6) = (m + 2) \times 2^{2m-6};$$

number 1s in the solutions of $g(x_1, x_2, \ldots, x_{2m}) = 0$

$\ldots\ldots$

for $(x_1, x_2, \ldots, x_{2m}) = (1, 1, \ldots, 1, 0, *, *)$, there are

$$(2m - 3) \times 2^2 + 2 \times 2 = (m + (m - 2)) \times 2^2$$

number 1s in the solutions of $g(x_1, x_2, \ldots, x_{2m}) = 0$; for $(x_1, x_2, \ldots, x_{2m}) = (1, 1, \ldots, 1, 0)$, there are $(2m - 1) \times 1$ number 1s in the solution of $g(x_1, x_2, \ldots, x_{2m}) = 0$.

Hence,

$$\begin{aligned}
c_{2m} &= \sum_{(x_1, x_2, \ldots, x_{2m}) \in D_{2m}(g)} (x_1 + x_2 + \cdots + x_{2m}) \\
&= m \times 2^{2m-2} + (m + 1) \times 2^{2m-4} + (m + 2) \times 2^{2m-6} \\
&\quad + \cdots + (m + m - 2) \times 2^2 + 2m - 1 \\
&= \frac{(3m + 1)4^{m} - (6m+1)}{9}.
\end{aligned}$$

(3)

In the same way, in the case of $n = 2m + 1$, we have

$$c_{2m+1} = \sum_{(x_1,x_2,\ldots,x_{2m+1})\in D_{2m+1}(g)} (x_1 + x_2 + \ldots + x_{2m+1})$$

$$= (2m + 1) \times 2^{2m-2} + (2m + 3) \times 2^{2m-4} + (2m + 5) \times 2^{2m-6} + \cdots$$

$$+ (2m + 2m - 1) \times 2^0 + 2m + 1$$

$$= \frac{(6m + 5)4^m + 6m + 4}{9}. \tag{4}$$

From ③, ④, we have

$$c_9 = 828 < 3986 = c_{11}, \quad c_{10} = 1817 < 8643 = c_{12}.$$

Therefore, the maximum n satisfying

$$\sum_{(x_1,x_2,\ldots,x_n)\in D_n(g)} (x_1 + x_2 + \cdots + x_n) \le 2017 \text{ is } n = 10.$$

$\square$

3 Given positive real numbers $a_1, a_2, \ldots, a_{n+1}$, prove

$$\sum_{i=1}^{n} a_i \cdot \sum_{i=1}^{n} a_{i+1} \ge \sum_{i=1}^{n} \frac{a_i a_{i+1}}{a_i + a_{i+1}} \cdot \sum_{i=1}^{n} (a_i + a_{i+1}).$$

Solution Since

$$\sum_{i=1}^{n} a_i \cdot \sum_{i=1}^{n} a_{i+1} = \left(\frac{\sum_{i=1}^{n} a_i + \sum_{i=1}^{n} a_{i+1}}{2}\right)^2 - \left(\frac{\sum_{i=1}^{n} a_{i+1} - \sum_{i=1}^{n} a_i}{2}\right)^2$$

$$= \frac{\left(\sum_{i=1}^{n}(a_i + a_{i+1})\right)^2}{4} - \left(\frac{a_{n+1} - a_1}{2}\right)^2, \quad \text{and}$$

$$4\sum_{i=1}^{n} \frac{a_i a_{i+1}}{a_i + a_{i+1}} = \sum_{i=1}^{n} \left[(a_i + a_{i+1}) - \frac{(a_i - a_{i+1})^2}{a_i + a_{i+1}}\right]$$

$$= \sum_{i=1}^{n}(a_i + a_{i+1}) - \sum_{i=1}^{n} \frac{(a_i - a_{i+1})^2}{a_i + a_{i+1}},$$

the original inequality is equivalent to

$$\frac{\left(\sum_{i=1}^{n}(a_i + a_{i+1})\right)^2}{4} - \frac{(a_{n+1} - a_1)^2}{4}$$

$$\geq \frac{1}{4}\left[\sum_{i=1}^{n}(a_i + a_{i+1}) - \sum_{i=1}^{n}\frac{(a_i - a_{i+1})^2}{a_i + a_{i+1}}\right] \cdot \sum_{i=1}^{n}(a_i + a_{i+1})$$

$$\Leftrightarrow \sum_{i=1}^{n}\frac{(a_i - a_{i+1})^2}{a_i + a_{i+1}} \cdot \sum_{i=1}^{n}(a_i + a_{i+1}) \geq (a_{n+1} - a_1)^2.$$

According to Cauchy's inequality, the final inequality holds, so the proposition is true. $\qquad\square$

4 For any positive integer n, let D_n be the set of the positive divisors of n, and $f_i(n)$ be the number of the elements in the following set

$$F_i(n) = \{a \in D_n \mid a \equiv i \pmod 4\} \quad (i = 0, 1, 2, 3).$$

Find the smallest poisitve integer m such that

$$f_0(m) + f_1(m) - f_2(m) - f_3(m) = 2017.$$

Solution Express positive integer m into the standard decomposition form $m = 2^{\alpha} \prod_{i=1}^{k} p_i^{\beta_i} \prod_{j=1}^{l} q_j^{\gamma_j}$, where α may be zero, $p_1, \ldots, p_k$ are the prime factors of m modulo 4 remainder 1, and $q_1, \ldots, q_l$ are the prime factors of m modulo 4 remainder 3 (if k or l is zero, the corresponding product is defined as 1). Let $P = \prod_{i=1}^{k}(1 + \beta_i)$, $Q = \prod_{j=1}^{l}(1 + \gamma_j)$. We will prove the following propositions:

(1) when $\alpha \geq 1$, $f_0(m) = (\alpha - 1)PQ$, $f_2(m) = PQ$;
(2) $f_1(m) + f_3(m) = PQ$;
(3) $f_1(m) = P \cdot \left\lceil\dfrac{Q}{2}\right\rceil$ (where $\left\lceil\dfrac{Q}{2}\right\rceil$ denotes the smallest integer not less than $\dfrac{Q}{2}$). Consider any positive factor of m, $u = 2^{\alpha'} \prod_{i=1}^{k} p_i^{\beta_i'} \prod_{j=1}^{l} q_j^{\gamma_j'}$, where

$$\alpha' \in \{0, 1, \ldots, \alpha\}, \beta_i' \in \{0, 1, \ldots, \beta_i\}(1 \leq i \leq k),$$
$$\gamma_j' \in \{0, 1, \ldots, \gamma_j\}(1 \leq j \leq l).$$

When $\alpha \geq 1$, $a \in F_0(m)$ if and only if $\alpha' \geq 2$, $a \in F_2(m)$ if and only if $\alpha' = 1$, and there are PQ ways to take the values of β_I', γ_j'. So Proposition (1) holds.

Since $f_1(m) + f_3(m)$ is the number of the odd factors of m, it is obviously equal to PQ. So Proposition (2) holds too.

For the proof of Proposition (3), note that

$$a \in F_1(m) \text{ if and only if } \alpha' = 0 \quad \text{and} \quad \sum_{j=1}^{l} \gamma_j' \equiv 0 \ (\mathrm{mod} \ 2).$$

We perform mathematical induction on l to prove that the number of qualified $(\gamma_1', \gamma_2', \ldots \gamma_l')$ is $\left\lceil \dfrac{Q}{2} \right\rceil$.

When $l = 0$, we have $Q = 1$, and $\left\lceil \dfrac{Q}{2} \right\rceil = 1$, so the proposition holds.

Assuming that the proposition holds for $l - 1$, consider the situation of l. There are the following two cases:

(a) If there is an odd number γ_j $(1 \leq j \leq l)$, we may assume that $j = l$. Then there are $(1 + \gamma_j)$ ways to take values of $\gamma_1', \ldots, \gamma_{l-1}'$, and for each of them there are $\dfrac{1 + \gamma_l}{2}$ ways to take a value of $\gamma_l' \in \{0, 1, \ldots, \gamma_l\}$ such that $\gamma_1' + \cdots + \gamma_{l-1}' + \gamma_l' \equiv 0 \ (\mathrm{mod} \ 2)$. Hence, the number of the qualified ways is

$$\frac{1 + \gamma_l}{2} \cdot \prod_{j=1}^{l-1} (1 + \gamma_j) = \frac{Q}{2} = \left\lceil \frac{Q}{2} \right\rceil.$$

(b) If all γ_j $(1 \leq j \leq l)$ are even numbers, then by induction we know, for $\gamma_l' = 0$, there are $\lambda_1 = \left\lceil \dfrac{1}{2} \prod_{j=1}^{l-1} (1 + \gamma_j) \right\rceil$ ways to take values of $\gamma_1', \ldots, \gamma_{l-1}'$ such that $\gamma_1' + \cdots + \gamma_{l-1}' \equiv 0 \ (\mathrm{mod} \ 2)$. As for $\gamma_l' \in \{1, 2, \ldots, \gamma_l\}$, by the same way as in (a), we can first take any values of $\gamma_1', \ldots, \gamma_{l-1}'$, and then take such a value of $\gamma_l' \in \{1, 2, \ldots, \gamma_l\}$ that satisfies $\gamma_1' + \cdots + \gamma_{l-1}' + \gamma_l' \equiv 0 \ (\mathrm{mod} \ 2)$ — in this case there are $\lambda_2 = \dfrac{\gamma_l}{2} \cdot \prod_{j=1}^{l-1} (1 + \gamma_j)$ ways. So there are totally $\lambda_1 + \lambda_2 = \left\lceil \dfrac{Q}{2} \right\rceil$ qualified ways.

Hence, by mathematical induction we know, for each l the number of the qualified ways to take values of $(\gamma_1', \gamma_2', \ldots, \gamma_l')$ is $\left\lceil \dfrac{Q}{2} \right\rceil$.

By the principle of multiplicaition we get

$$f_1(m) = |F_1(m)| = 1 \cdot \prod_{i=1}^{k} (1 + \beta_i) \cdot \left\lceil \frac{Q}{2} \right\rceil = P \cdot \left\lceil \frac{Q}{2} \right\rceil.$$

Proposition (3) is proved.

Now suppose positive integer m satisfies

$$f_0(m) + f_1(m) - f_2(m) - f_3(m) = 2017.$$

If m is an odd number, then $f_0(m) = f_2(m) = 0$. By Propositions (2), (3) we have

$$2017 = f_1(m) - f_3(m) = 2f_1(m) - PQ = P \cdot \left[2 \left\lceil \frac{Q}{2} \right\rceil - Q \right].$$

Q must be an odd number, and then $\left\lceil \dfrac{Q}{2} \right\rceil - Q = 1$. Hence $P = \prod_{i=1}^{k} (1+\beta_i) = 2017$. Since 2017 is prime number, we have $k = 1$, $\beta_1 = 2016$. Then

$$m = p_1^{2016} \prod_{j=1}^{l} q_j^{\gamma_j} \geq 5^{2016}.$$

If m is an even number, then by Propositions (1), (2), (3) we have

$$2017 = f_0(m) - f_2(m) - (f_1(m) + f_3(m)) + 2f_1(m)$$

$$= (\alpha - 1)PQ - PQ - PQ + 2P \cdot \left\lceil \frac{Q}{2} \right\rceil = P \cdot \left[(\alpha - 3)Q + 2 \cdot \left\lceil \frac{Q}{2} \right\rceil \right]. \tag{1}$$

When $P = 2017$, we have $m \geq 2p_1^{2016} > 5^{2016}$.

If $P = 1$, $(\alpha - 3)Q + 2 \cdot \left\lceil \dfrac{Q}{2} \right\rceil = 2017$. $(\alpha - 3)Q$ is obviously an odd number, and so is Q. Hence, $(\alpha - 2)Q = 2016 = 2^5 \times 3^2 \times 7$, where $2^5 \mid \alpha - 2$. If $\alpha - 2 > 2^5$, then $\alpha - 2 \geq 2^5 \times 3$, and $m \geq 2^{98}$. If $\alpha - 2 = 2^5$, since $Q = 7 \times 3 \times 3 = 9 \times 7 = 21 \times 3 = 63$, we need to handle the following 4 cases:

(a) $l = 3$, so the values of $1 + \gamma_1$, $1 + \gamma_2$, $1 + \gamma_3$ are a permutation of 7, 3, 3. As q_1, q_2, q_3 are different prime numbers modulo 4 remainder 3, we have

$$m \geq 2^{34} \times 3^6 \times 7^2 \times 11^2.$$

(b) $l = 2$, and $\{1 + \gamma_1, 1 + \gamma_2\} = \{9, 7\}$. At this time, we have

$$m \geq 2^{34} \times 3^8 \times 7^6 > 2^{34} \times 3^6 \times 7^2 \times 11^2.$$

(c) $l = 2$, and $\{1 + \gamma_1, 1 + \gamma_2\} = \{21, 3\}$. We have

$$m \geq 2^{34} \times 3^{20} \times 7^2 > 2^{34} \times 3^6 \times 7^2 \times 11^2.$$

(d) $l = 1$, and $1 + \gamma_1 = 63$. We have

$$m \geq 2^{34} \times 3^{62} > 2^{34} \times 3^6 \times 7^2 \times 11^2.$$

In summary, noting $2^{34} \times 3^6 \times 7^2 \times 11^2 < 2^{34} \times 16^{10} < 2^{98}$, we have

$$m \geq \min\{5^{2016}, 2^{34} \times 3^6 \times 7^2 \times 11^2\} = 2^{34} \times 3^6 \times 7^2 \times 11^2.$$

On the other hand, if $m = 2^{34} \times 3^6 \times 7^2 \times 11^2$, then m is an even number, and $\alpha = 34$, $P = 1$, $Q = 63$, satisfying Equation ①. Therefore, the required smallest positive integer is $m = 2^{34} \times 3^6 \times 7^2 \times 11^2$. □

Second Day
(8:00 – 12:00; July 31, 2017)

⑤ Let a, b, c be real numbers, and $a \neq 0$. If equation $2ax^2 + bx + c = 0$ has a root in $[-1, 1]$, prove:

$$\min\{c, a + c + 1\} \leq \max\{|b - a + 1|, |b + a - 1|\},$$

and determine the necessary and sufficient conditions for a, b, c when the the equality holds.

Solution Noting $\max\{|b - a + 1|, |b + a - 1|\} = |b| + |a - 1|$, the original inquality is equivalent to

$$\min\{c, a + c + 1\} \leq |b| + |a - 1|. \tag{1}$$

By the given condition we know there is $x_0 \in [-1, 1]$, satisfying $2ax_0^2 + bx_0 + c = 0$. We will discuss the problem in the following 2 situations.

(1) If $a > 0$, then

$$\min\{c, a + c + 1\} = c = -2ax_0^2 - bx_0 \leq -bx_0$$

$$\leq |b| \leq |b| + |a - 1|.$$

Hence, Inequality ① holds, and the equality holds if and only if $|a - 1| = 0$, $-2ax_0^2 = 0$, $c = -bx_0 = |b|$, i.e., $a = 1$, $x_0 = 0$, $c = b = 0$.

(2) If $a < 0$, then

$$\min\{c, a + c + 1\} \leq a + c + 1 = a - 2ax_0^2 - bx_0 + 1$$
$$\leq a + (-2a) + |b| + 1 = |b| + 1 - a = |b| + |a - 1|.$$

Inequality ① still holds, and the equality holds if and only if $a + c + 1 \leq c$, $-2ax_0^2 = -2a$, $-bx_0 = |b|$, from which we get $a \leq -1$, $x_0^2 = 1$, and x_0 and b are in opposite signs. Therefore,

$$2a - |b| + c = 2ax_0^2 + bx_0 + c = 0.$$

By (1) and (2) we knows Inequatlity ① holds, and the necessary and sufficient conditions for a, b, c to reach the equality in ① is $(a, b, c) = (1, 0, 0)$, or $a \leq -1$, $2a - |b| + c = 0$.

Remark   We can also prove the inequality in the following way: Suppose the equation has a root $x_0 \in [-1, 1]$, satisfying $2ax_0^2 + bx_0 + c = 0$, or

$$(a + 1)x_0^2 + c = -[(a - 1)x_0^2 + bx_0] = -x_0 \cdot [(a - 1)x_0 + b].$$

Then

$$\min\{c, a + c + 1\} \leq (a + 1)x_0^2 + c \leq |x_0| \cdot |(a - 1)x_0 + b|$$
$$\leq |(a - 1)x_0 + b| \leq \max\{|a - 1 + b|, |-a + 1 + b|\}.$$

Based on the above formulas, we can also obtain the necessary and sufficient conditions for reaching the equality in ①.

6   As shown in Fig. 6.1, in the inscribed quadrilateral $ABCD$ of circle O, the diagonals AC and BD are perpendicular to each other, M is the midpoint of arc $\overset{\frown}{ADC}$, and the circle passing through M, O, D intersects with DA and DC at points E and F, respectively.

Proof $BE = BF$

Solution   As shown in Fig. 6.2, connecting ME, MF, MD, MA, MB, MC, EF. Since M, E, O, F, D are four concyclic points, we have

$$\angle MFE = \angle MDE = \angle MDA = \angle MCA,$$

$$\angle MEF = 180° - \angle MDF = 180° - \angle MDC = \angle MAC.$$

Furthermore, $\angle MCA = \angle MAC$, because M is the midpoint of $\overset{\frown}{ADC}$. Hence, $\angle MFE = \angle MEF$, and then $ME = MF$.

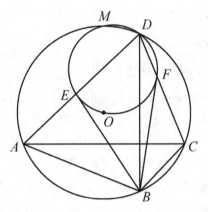

Fig. 6.1

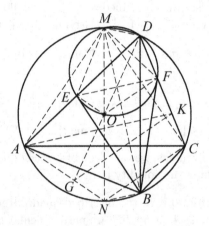

Fig. 6.2

Let N be another point at wich line MO and circle O intersect. Connecting OE, OF, NA, NC, we have

$$\angle DEO = \angle DMO = \angle DMN = \angle DAN,$$

Hence, $OE//AN$. In the same way, $OF//NC$.

Connect DO and extend it to intersect AN at point G. Pass G to make $GK//NC$ and intersect DC at K. Then $\triangle AGK$ and $\triangle EOF$ are homothetic triangles with D as their homothetic center, so $EF//AK$.

Since $GK//NC$ and A, N, C, D are four concyclic points, we have

$$\angle AGK = \angle ANC = 180° - \angle ADC.$$

Therefore, A, G, K, D are concyclic and $\angle AKG = \angle ADG$. Furthermore, as $DB \perp AC$, then $\angle BDC = 90° - \angle ACD = \angle ADG$. Hence,

$$\angle BNC = \angle BDC = \angle ADG = \angle AKG.$$

Combing with $GK//NC$, we get $NB//AK$. Furthermore, as MN is the diameter of circle O, we know $BM \perp NB$, and then $BM \perp AK$. Hence, $BM \perp EF$. Therefore, BM is the vertical bisector of EF, so $BE = BF$. $\qquad\square$

7 Find the largest positive integer$n$, such that there are mutually different positive integers $x_1, x_2, \ldots, x_n$ satisfying

$$x_1^2 + x_2^2 + \cdots + x_n^2 = 2017.$$

Solution Since the sum of squares of the smallest 18 mutually different positive integers is $1^2 + 2^2 + \cdots + 18^2 = 2109 > 2017$, then $n \leq 17$.

We will prove by contradiction that $n \neq 17$. Assume there are 17 mutually different positive integers $x_1, x_2, \ldots, x_{17}$, satisfying

$$x_1^2 + x_2^2 + \cdots + x_{17}^2 = 2017. \qquad \text{(1)}$$

It is easy to see that 2017 is a prime of type $3N + 1$, and if x_j is not a multiple of 3, $x_j^2 \equiv 1 \pmod 3$. So there are $3a + 1$ numbers in $x_1, x_2, \ldots, x_{17}$ that are not the multiples of 3, and the rest $3b + 1$ ones are divisible by 3, where

$$a, b \in \mathbb{N}, \quad \text{and} \quad (3a + 1) + (3b + 1) = 17, \quad \text{or} \quad a + b = 5. \qquad \text{(2)}$$

Note that $1^2 + 2^2 + \cdots + 17^2 = 1785$, and $2017 - 1785 = 232$; there are 5 numbers in $M = \{1, 2, \ldots, 17\}$ that are divisible by 3, while the rest 12 ones are coprime with 3. Hence, in order that Equation (1) has a solution, it is necessary to replace k numbers in M with another k ones that are greater than 17, so that the new 17 numbers satisfy Condition (2) and their sum of squares increases by 232.

Assume $k = 1$. It is obvious that we need to remove a multiple of 3 from M and add a number y that is not divisible by 3 and satisfies

$$(3x)^2 + 232 = y^2, \quad x \in \{1, 2, 3, 4, 5\}, \quad y > 17.$$

It is impossible.

Assume $k = 2$. Then we have the following 3 cases.

(1) Remove 2 numbers from M that are not divisible by 3, and add 2 ones that are divisible by 3 and great than 17. First consider removing the

two largest non-3 multiples in M, and adding two minimum 3 multiples greater than 17. We get

$$(18^2 + 21^2) - (16^2 + 17^2) = 220 \neq 232.$$

It is not a solution of the problem. Next, when at least one of the numbers removed from M is replaced with a smaller number, or at least one of the numbers added from outside M is replaced with a larger number, the above difference will be greater than 232. They are also not solutions.

(2) Remove a multiple of 3 and a non-multiple of 3 from M, and add two non-multiples of 3 outside of M. Similarly, consider the largest number removed in M and the smallest number added outside of M. We get

$$(19^2 + 20^2) - (15^2 + 17^2) = 247 > 232.$$

It is also not a solution.

(3) Remove two multiples of 3 from M, and add a multiple of 3 and a non-multiple of 3 outside of M. Similarly, consider the largest qualified number removed from M and the smallest qualified number added outside of M. We get

$$(18^2 + 19^2) - (12^2 + 15^2) = 316 > 232.$$

It is also not a solution.

Assume $k \geq 3$. Then $a_1, \ldots, a_k$, are removed from M and $b_1, \ldots, b_k$, are added outside M. It is obvious that

$$(b_1^2 + b_2^2 + \cdots + b_k^2) - (a_1^2 + a_2^2 + \cdots + a_k^2)$$
$$\geq (18^2 + 19^2 + 20^2) - (15^2 + 16^2 + 17^2)$$
$$= 315 > 232,$$

It is also not a solution.

Therefore, $n \leq 16$. On the other hand, when $n = 16$, we have

$$1^2 + 2^2 + \cdots + 16^2 = 1496 \quad \text{and} \quad 2017 - 1496 = 521.$$

As $(17^2 + 18^2 + 23^2) - (13^2 + 14^2 + 16^2) = 521$, replacing 13, 14, 16 with 17, 18, 23, we can see that $1, 2, 3, 4, 5, 6, 7, 8, 9, 10, 11, 12, 15, 17, 18, 23$ are 16 mutually different positive integers and their sum of squares are 2017.

Therefore, the required largest positive integer is $n = 16$.

Remark The elements in set $X = \{x_1, x_2, \ldots, x_{16} \mid x_1^2 + x_2^2 + \cdots + x_{16}^2 = 2017\}$ are

$$X_1 = \{1, 2, 3, 4, 5, 6, 7, 8, 9, 10, 11, 12, 15, 17, 18, 23\},$$

$$X_2 = \{1, 2, 3, 4, 5, 6, 7, 8, 9, 10, 11, 13, 14, 16, 19, 23\},$$

$$X_3 = \{1, 2, 3, 4, 5, 6, 7, 8, 9, 11, 13, 14, 15, 16, 18, 21\},$$

$$X_4 = \{1, 2, 3, 4, 5, 6, 7, 8, 10, 11, 12, 13, 15, 17, 18, 21\},$$

$$X_5 = \{1, 2, 3, 4, 5, 6, 7, 9, 10, 11, 12, 13, 14, 15, 20, 21\},$$

$$X_6 = \{1, 2, 3, 4, 6, 7, 8, 9, 10, 11, 12, 13, 14, 15, 19, 21\},$$

$$X_7 = \{1, 2, 3, 4, 5, 6, 7, 9, 11, 12, 13, 14, 15, 16, 18, 19\},$$

$$X_8 = \{2, 3, 4, 5, 6, 7, 8, 9, 10, 11, 12, 13, 15, 17, 18, 19\}.$$
□

8 Given integers $m \geq 3$, $n \geq 3$, let $S = \{(a, b) \mid a \in \{1, 2, \ldots, m\},$ $b \in \{1, 2, \ldots, n\}\}$, and A be subset of S. Suppose there are not positive integers x_1, x_2, x_3, y_1, y_2, y_3, satisfying $x_1 < x_2 < x_3$, $y_1 < y_2 < y_3$, and $(x_1, y_2), (x_2, y_1), (x_2, y_2), (x_2, y_3), (x_3, y_2) \in A$. Find the maximum number of the elements in A.

Solution If there is a point (x_2, y_2) in a set X such that there are still points $(x_2, y_3), (x_2, y_1), (x_1, y_2), (x_3, y_2)$ in X that are above, below, to the right of, and to the left of (x_1, y_2), respectively, X is then called a "central point set". Hence, the problem is equivalent to determining how many points can be taken out of S so that the set A composed of these points is not a central point set.

Firstly, we take all the points with the abscissa 1 or m and the points with the ordinate 1 or n from S, to form set A_0. Obviously, A_0 is not a center point set, and we have $|A_0| = 2m + 2n - 4$.

Next, we will prove: if $A \subseteq S$, $|A| \geq 2m + 2n - 3$, then A is a central point set.

For $1 \leq i \leq n$, suppose there are k_i points in A with the ordinate i. If $k_i \geq 3$, we color the points red in row i except for the rightmost and leftmost ones. Then there are at least $(k_1 - 2) + \cdots + (k_n - 2)) = |A| - 2n$ red points in A.

For $2 \leq j \leq m - 1$, suppose there are l_j points in A with the abscissa j. If $l_j \geq 3$, we color the points blue in column j except for the top and bottom ones, and also color all the points blue with the abscissa 1 or m.

Then there are at least $l_1 + (l_2 - 2) + \cdots + (l_{m-1} - 2) + l_m = |A| - 2m + 4$ blue points in A.

Note that

$$(|A| - 2n) + (|A| - 2m + 4) = (|A| - 2n - 2m + 3) + |A| + 1 > |A|.$$

Hence, there is a point in A being colored both red and blue: it is easy to check that there are other points in A which are above below, to the right of and to the left of this point, respectively. Therefore, A is a central point set.

In summary, the maximum number of elements in A is $2m + 2n - 4$.

Remark We can also regard S as an $n \times m$ matrix S_{nm}. Then if there is a row or a column in S_{nm} that contains at most 2 points of A, we remove this row or colun from S_{nm}, to consider the remaining $S_{n-1,m}$ or S_{nm-1}. So we only need to consider the situation where each row or column of S_{nm} contains at least 3 points in A. At this time we still need the condition that $|A| \geq 2m + 2n - 3$. Otherwise, the following figure shows that A may not be a central point set.

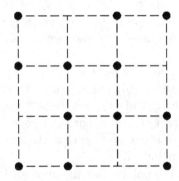

International Mathematical Olympiad

2017 (Rio de Janeiro, Brazil)

The 58th International Mathematical Olympiad (IMO) was held in Rio de Janeiro, Brazil from July 12 to 23, 2017, Brazilian time. 615 students from 111 countries and regions participated in the competition. The Chinese team won second place in the group with a total of 159 points. 6 players won 5 gold medals and 1 silver medal.

The members of the Chinese team are as follows:

Team Leader: Yao Yijun, Fudan University

Deputy Leader: Zhang Sihui, University of Shanghai for Science and Technology

Observer: Qu Zhenhua, East China Normal University;

Wan Jun, Shanghai Middle School Affiliated to Fudan University

Team Contestants:

Zhou Xingjian, Middle School Affiliated to Renmin University of China; 23 points, Silver Medal

He Tiancheng, Middle School Affiliated to South China Normal University; 25 points, Gold Medal

Ren Qiuyu, High School Affiliated to South China Normal University; 32 points, Gold Medal

Wu Jinze, Wuhan No.2 Middle School of Hubei Province; 26 points, Gold Medal

Zhang Lu, Changjun Middle School, Changsha City, Hunan Province; 28 points, Gold Medal

Jiang Yunyang, Yinzhou Middle School of Ningbo City, Zhejiang Province; 25 points, Gold Medal

The top ten teams and their total scores are as follows:

1. South Korea;   170 points
2. China;   159 points
3. Vietnam;   155 points
4. United States;   148 points
5. Iran;   142 points
6. Japan;   134 points
7. Singapore;   131 points
8. Thailand;   131 points
9. Taiwan;   130 points
10. UK   130 points

The gold medal line is 25 points, the silver medal line is 19 points, and the bronze medal line is 16 points.

The Six contestants of the Chinese National Team received training at High School Affiliated to Fudan University (HSAFU), Shanghai, from June 21 to July 4. Yao Yijun, Chen Xiaomin, Qu Zhenhua, Zhang Sihui, He Yijie, Wang Guozhen, Wang Shanwen, Jiang Zilin, and Zhang Ruixiang provided tutoring to them. Wan Jun from HSAFU acted as the head teacher, Wu Jian, Principal of HSAFU, and Chen Jinhui, Director of the HSAFU Affairs Office, did a lot of work for the team.

From July 9th to 14th, the national team conducted training and preparations before departure at the School of Mathematics, Peking University. Yuan Hanhui and Wang Bin provided guidance to the contestants. Professor Chen Dayue, Secretary General of the Chinese Mathematical Society and Dean of the School of Mathematical Sciences, Professor Liu Bin, former vice dean of the School of Mathematical Sciences, Peking University, Professor Li Ruo, deputy dean of teaching of the School of Mathematical Sciences, Peking University, Professor Wu Jianping, deputy director of the Popularization Committee of the Chinese Mathematical Society and the Olympic Committee, had a discussion with the contestants. Professor Qiu Zonghu met the contestants and provided pre-match tutoring. Academician Tian Gang, vice president of Peking University, gave great support to the training, and Sun Zhaojun, deputy secretary of the party committee of the School of Mathematical Sciences of Peking University, did a lot of work for the national team.

First Day
(9:00 – 13:00; July 18, 2017)

1 For each integer $a_0 > 1$, define the sequence $a_0, a_1, a_2, \ldots$ by:

$$a_{n+1} = \begin{cases} \sqrt{a_n} & \text{if } \sqrt{a_n} \text{ is an integer,} \\ a_n + 3 & \text{otherwise,} \end{cases} \quad \text{for each } n > 0.$$

Determine all values of a_0 for which there is a number A such that $a_n = A$ for infinitely many values of n. (Contributed by South Africa)

Solution If and only if a_0 is a multiple of all 3, it satisfies the given condition.

Since a_{n+1} is determined only by a_n, there are infinite terms with the same value in $\{a_n\}$ if and only if the sequence is finally periodic and, equivalently, bounded. Note that the modulus 3 remainder of a square number is always either 0 or 1, so if $a_k \equiv 2 \pmod 3$, then a_k is not a square number, and by recursion we know, for $m \geq k$, $a_{m+1} = a_m + 3 \equiv 2 \pmod 3$; therefore, $\{a_n\}$ is strictly increasing after the k term, without upper bound. In particular, if $a_0 \equiv 2 \pmod 3$, then a_0 cannot satisfy the given condition.

If $3 \mid a_k$, we always have $3 \mid a_{k+1}$ no matter $a_{k+1} = \sqrt{a_k}$ or $a_{k+1} = a_k + 3$. In particular, when $3 \mid a_0$, every term in $\{a_n\}$ can be divided by 3. At this time, let a square number $N^2 > a_0$, satisfying $3 \mid N$. We show that for any positive integer n, $a_n \leq N^2$. By reduction to absurdity, assume there is the smallest positive integer k, such that $a_k > N^2$. Since $3 \mid a_k$, we have $a_k \geq N^2 + 3$, $a_{k-1} \leq N^2$, and $a_k - a_{k-1} \leq 3$. Then $a_{k-1} = N^2$, but by the definition we have $a_k = N$, a contradiction.

Therefore, when $3 \mid a_0$, $\{a_n\}$ is bounded. So a_0 satisfies the given condition.

Finally consider the case $a_0 \equiv 1 \pmod 3$. It is easy to see that if $3 \nmid a_k$ and $a_k > 1$, then $3 \nmid a_{k+1}$ and $a_{k+1} > 1$. Therefore, every term in $\{a_n\}$ cannot be divided by 3 and is greater than 1. If at this time $\{a_n\}$ is bounded, then every term is of remainder 1 modulous 3, and there are some values that are taken by infinite many terms. Let A be the largest value that is taken by infinite many terms in $\{a_n\}$. Then A is a square number. Otherwise, assume $a_k = A$. We have $a_{k+1} = a_k + 3 = A + 3 > A$. The value $A + 3$ is obviously also taken by infinite many terms in $\{a_n\}$, but it is impossible by the definition of A. So there are integers N, such that $A = N^2$ and $a_k = N^2$. Then $a_{k+1} = N$. Therefore $N \equiv 1 \,(\mathrm{mod}\,3)$ and $N > 1$, i.e. $N \geq 4$. There exists $j > k$, such that $a_j = a_k = N^2 > (N-2)^2$, while $a_{k+1} = N \leq (N-2)^2$ (since $N \geq 4$). So there exists the smallest number $l > k+1$, such that $a_l > (N-2)^2$. Then $a_{l-1} \leq (N-2)^2 < a_l$, and $a_l = a_{l-1} + 3$. But from $a_l \equiv a_{l-1} \equiv 1 \,(\mathrm{mod}\,3)$, we get $a_{l-1} = (N-2)^2 \equiv 1 \,(\mathrm{mod}\,3)$. Therefore $a_l = N - 2$, a contradiction.

Therefore, $a_0 \equiv 1 \,(\mathrm{mod}\,3)$ cannot satisfy the given condition too. The proof is completed.

Remark   This is a simple problem in number theory. Basically all solutions need to discuss several cases separately according to the congruence class of a_0 modulo 3. In the scoring criteria, a detailed description of the incomplete solution is given below

- No reason is given, only the answer or intermediate conclusion is written, or only a few cases about a_0 are discussed. 0 points.
- Only give a complete answer to $a_0 \equiv 0 \,(\mathrm{mod}\,3)$. 3 points.
- Only give a complete answer to $a_0 \equiv 1 \,(\mathrm{mod}\,3)$. 4 points.
- Only give a complete answer to $a_0 \equiv 2 \,(\mathrm{mod}\,3)$. 1 point.
- Only give a complete answer to $a_0 \equiv 0 \,(\mathrm{mod}\,3)$ and $a_0 \equiv 2 \,(\mathrm{mod}\,3)$. 4 points.
- Only give a complete answer to $a_0 \equiv 1 \,(\mathrm{mod}\,3)$ and $a_0 \equiv 2 \,(\mathrm{mod}\,3)$. 5 points.
- Only give a complete answer to $a_0 \equiv 0 \,(\mathrm{mod}\,3)$ and $a_0 \equiv 1 \,(\mathrm{mod}\,3)$. 4 points.
- If there is a small error (may be repeated several times), such as the induction base is not written clearly, or the calculation is wrong, 1 point is deducted.
- If there are two small errors of different kinds, for example, the induction base is not written clearly, and the calculation is wrong, a maximum of 2 points will be deducted.

2   Let $\mathbb{R}$ be the set of real numbers. Determine all functions $f\colon \mathbb{R} \to \mathbb{R}$ such that, for all real numbers x and y, $f(f(x)f(y))+f(x+y) = f(xy)$. (Contributed by Albania)

Solution 1 First, we notice that if f is a solution of the given functional equation, then $-f$ is also a solution of it, so we might as well set $f(0) \leq 0$.

Let $(x, y) = (0, 0)$, we have

$$f(f(0)^2) = 0. \tag{1}$$

For any real number $x \neq 1$, let $y = \dfrac{x}{x-1}$, so that $x + y = xy$. Then

$$f\left(f(x) \cdot f\left(\frac{x}{x-1}\right)\right) = 0, \quad x \neq 1. \tag{2}$$

Now we discuss the following two cases about $f(0)$.

Case 1: $f(0) = 0$. At this time we get $f(x) = 0$ for any real number x. Therefore, f is a constant with value 0.

Case 2: $f(0) < 0$.

Claim 1 $f(a) = 0$ *if and only if* $a = 1$.

Proof Otherwise, from ①, there exists $a > 0$ such that $f(a) = 0$. If $a \neq 1$, in ② let $x = a$, and we get $f(0) = 0$, contradicting $f(0) \neq 0$.

From ① we have $f(0)^2 = 1$, so $f(0) = -1$. In the original equation let $y = 1$, we get

$$f(0) + f(x+1) = f(x),$$

i.e., $f(x+1) = f(x)+1$. By mathematical induction we get: for any integer n and real number x,

$$f(x + n) = f(x) + n. \tag{3}$$

Claim 2 f *is an injective function*.

Proof By reduction to absurdity, assume there are real numbers $a, b, a \neq b$, satisfying $f(a) = f(b)$. For any integer N, from ③ we have

$$f(a + N + 1) = f(b + N) + 1.$$

Let $N > b$. It is easy to find, there are real numbers x_0, y_0, satisfying $x_0 + y_0 = a + N + 1$, $x_0 y_0 = b + N$. Since $a \neq b$, then $x_0 \neq 1$, $y_0 \neq 1$. Let $x = x_0 y = y_0$. The original equation becomes

$$f(f(x_0)f(y_0)) + f(a + N + 1) = f(b + N).$$

That is, $f(f(x_0)f(y_0)) + 1 = f(f(x_0)f(y_0) + 1) = 0$. By Claim 1, we get $f(x_0)f(y_0) + 1 = 1$, i.e., $f(x_0)f(y_0) = 0$. But this is impossible, since $x_0 \neq 1$, $y_0 \neq 1$.

Now let $x = t, y = -t$. The original equation becomes

$$f(f(t)f(-t)) + f(0) = f(-t^2).$$

That is, $f(f(t)f(-t)) = f(-t^2) + 1 = f(-t^2 + 1)$. Since f is injectve, we have

$$f(t)f(-t) = -t^2 + 1. \tag{4}$$

Let $x = t$, $y = 1 - t$. The original equation becomes

$$f(f(t)f(1 - t)) + f(1) = f(t(1 - t)).$$

As $f(1) = 0$, we have $f(f(t)f(1 - t)) = f(t(1 - t))$, and by Claim 2,

$$f(t)f(1 - t) = t(1 - t). \tag{5}$$

Because $f(1 - t) = 1 + f(-t)$, from ④, ⑤ we get $f(t) = t - 1$. If $f(0) = 1$, then $f(t) = 1 - t$.

It is easy to check, $f_1(x) = 0$, $f_2(x) = x - 1$, and $f_3(x) = 1 - x$ all satisfy the requirements.

Solution 2 After Claim 1 in Solution 1, we may not prove Claims 2, but go in the following way: Deifne $g(x) = f(x + 1) = f(x) + 1$, and in the original functional equantion substitute x and y by $x+1$ and $y+1$, respectively. We have

$$f(f(x + 1)f(y + 1)) + f(x + y + 2) = f(xy + x + y + 1).$$

That is

$$g(g(x)g(y)) + g(x + y) = g(xy + x + y). \tag{6}$$

By Claim 1, we know 0 is the only zero of g. Now we prove $g(x) = x$.
Claim 3 Let $n \in \mathbb{Z}$, $x \in \mathbb{R}$. We have
(a) $g(x + n) = g(x) + n$, and $g(x) = n \Leftrightarrow x = n$.
(b) $g(nx) = ng(x)$.

Proof (a) From ③ , we get $g(x + n) = g(x) + n$, and
$$g(x) = n \Leftrightarrow g(x - n) = 0 \Leftrightarrow x - n = 0.$$

(b) It is obviously true when $x = 0$. When $x \neq 0$, in ⑥ let $y = \dfrac{n}{x}$. We have

$$g\left(g(x)g\left(\frac{n}{x}\right)\right) + g\left(x + \frac{n}{x}\right) = g\left(n + x + \frac{n}{x}\right)$$

$$\Leftrightarrow g\left(g(x)g\left(\frac{n}{x}\right)\right) = n$$

$$\Leftrightarrow g(x)g\left(\frac{n}{x}\right) = n.$$

Then $g(x) = \dfrac{n}{g\left(\dfrac{n}{x}\right)}$. Letting $n = 1$, $g(1/x) = 1/g(x)$. Substituting x with nx, we get

$$g(nx) = \frac{n}{g\left(\dfrac{1}{x}\right)} = ng(x).$$

Therefore, Claim 3(b) is true.

Claim 4 *For any real numbers a, b, we have*
$$g(a + b) = g(a) + g(b).$$

Proof Letting $n = -1$, from Claim 3(b) we know g is an odd function. In ⑥ substituting x and y with $-x$ and $-y$, respectively, we have
$$g(g(x)g(y)) - g(x + y) = -g(-xy + x + y).$$

Subracting it from ⑥, then
$$2g(x + y) = g(xy + x + y) + g(-xy + x + y).$$

Let $\alpha = xy + x + y$, $\beta = -xy + x + y$ and $n = 2$. Then by Claim 3(b) we have
$$g(\alpha + \beta) = g(\alpha) + g(\beta).$$

Note that for any real numbers α, β, satisfying
$$\left(\frac{\alpha + \beta}{2}\right)^2 - 4 \cdot \frac{\alpha - \beta}{2} \geq 0,$$

we can find x, y such that
$$x + y = \frac{\alpha + \beta}{2}, xy = \frac{\alpha - \beta}{2}.$$

Then for any real numbers a, b, we can let either $n = 1$ or $n = -1$ such that
$$g(na) + g(nb) = g(na + nb) \Leftrightarrow ng(a) + ng(b) = ng(a + b).$$

So Claim 4 is true.

Now let $y = 1$ in ⑥. Then by Claim 3, we have

$$g(g(x)g(1)) + g(x+1) = g(2x+1)$$
$$\Leftrightarrow g(g(x)) + g(x) + 1 = 2g(x) + 1$$
$$\Leftrightarrow g(g(x)) = g(x).$$

By Claim 4, $g(g(x) - x) = 0$, and by Claim 1, $g(x) - x = 0$. Therefore, $g(x) = x$.

Remark Various solutions to this question can be basically divided into two parts, so the scoring standards are also made according to them. The first part has a maximum of 3 points (the following three subparts can be accumulated):

- Write out at least one of the three answers: $f(x) = 0$, $f(x) = x - 1$, and $f(x) = 1 - x$. Get 1 point. If there are a limited number of wrong solutions on the answer sheet, no points will be deducted.
- Prove that if the function f has zeros not equal to 1, then $f \equiv 0$. Get 1 point.
- Prove that $f(0) = \pm 1$ and $f(x+1) = f(x) - f(0)$ for any real number x (or something like $f(x+n) = f(x) - nf(0)$, $n \in \mathbb{Z}$). Get 1 point.

The second part scores up to 4 points:

- Prove that f is injective (Claim 2 in Solution 1); get 3 points. Base on it, find all the solutions to the original function equation. Get another 1 point.
- In Solution 2, prove that g is an additive function. Get 3 points. On this basis, prove that we have found all the solutions to the original function equation. Get one more point.
- If you can also prove that the value range of f is a subfield of the real number field, which is closed for subtraction, get 1 point. If you further prove that it is closed for reciprocal, then get one more point. For the case of $f(0) = 1$, if you prove, $f(x) = 1 - x$, for x in the range, and infer, from the original function equation,

$$1 - f(x)f(y) + f(x+y) = f(xy),$$

for any real number x, y, then get one more point. Finally, if you, base on the result above, give the definition of $g(x) = f(x+1)$, prove g is an additive function, and find all the solutions of the original functional equation, you get still another 1 point.

Conclusions that clearly cannot be scored include:

- Only proved that $f(f(0)^2) = 0$.
- It is only observed that if f is a solution, then $-f$ is also a solution, without giving any non-zero solution.
- It is only proved that if $f(0) = 0$, f is always equal to 0, without further conclusions.

3 A hunter and an invisible rabbit play a game in the Euclidean plane. The rabbit's starting point, A_0, and the hunter's starting point, B_0, are the same. After $n - 1$ rounds of the game, the rabbit is at point A_{n-1} and the hunter is at point B_{n-1}. In the nth round of the game, three things occur in order.

- (i) The rabbit moves invisibly to a point A_n such that the distance between A_{n-1} and A_n is exactly 1.
- (ii) A tracking device reports a point P_n to the hunter. The only guarantee provided by the tracking device to the hunter is that the distance between P_n and A_n is at most 1.
- (iii) The hunter moves visibly to a point B_n such that the distance between B_n and B_{n-1} is exactly 1.

Is it always possible, no matter how the rabbit moves, and no matter what points are reported by the tracking device, for the hunter to choose her moves so that after 10^9 rounds she can ensure that the distance between her and the rabbit is at most 100? (Contributed by Austria)

Solution The hunters cannot guarantee that after 10^9 rounds the distance between her and the rabbit is at most 100.

First, if the solution to the problem is "the hunter has a strategy that meets the requirements of the problem", then this strategy should be applicable to any sequence of P_n (and any sequence of A_n that satisfies Condition (ii)). We will explain that if the hunter has a "bad luck", she cannot guarantee this.

Let $d_n = A_n B_n$. Suppose there exists $n \le 10^9$ such that $d_n \ge 100$. Then after this, the rabbit just needs to jump along the line between it and the hunter, in the opposite direction to the latter.

Now suppose $d_n < 100$. We prove that when the rabbit moves in a appropriate way and the feedback points of the positioning device are "beneficial" for the rabbit, no matter what kind of movement the hunter takes, after 200 rounds, there is no guarantee that $d_{n+200}^2 < d_n^2 + \dfrac{1}{2}$, or in other words, it is always possible that $d_{n+200}^2 \geq d_n^2 + \dfrac{1}{2}$. Then for $d_0 = 0$ at the beginning, after $n_0 = 2 \cdot 10^4 \cdot 200 = 4 \cdot 10^6 < 10^9$ rounds, the rabbit was able to successfully increase the distance between it and the hunter to at least 100.

Assume that after the nth round, the rabbit is at A_n and the hunter is at B_n. We can even let the rabbit show its position to the hunter (this makes the previous feedback information from the positioning device negligible). Let l be the straight line $A_n B_n$, and Y_1, Y_2 be two points on both sides of l, whose distances to A_n be both 200 (their distances to B_n be greater than 200) and that to l be both 1. As shown in the figure.

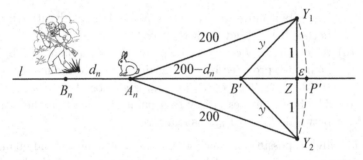

The rabbit's strategy is to choose one of Y_1 and Y_2 as the target and jump towards it along the straight line linking it and A_n. At the same time, suppose the points returned by the positioning device all fall on line l. In this way, the hunter will not know whether the rabbit jumped to Y_1 or Y_2.

In other words, we assume that the points of the next 200 positioning device feedbacks are exactly 200 points that advance a distance of 1 each time in the direction of the ray on l starting from A_n, that is,

$$A_n P_{n+1} = P_{n+1} P_{n+2} = \ldots = P_{n+199} P_{n+200} = 1.$$

Denote $P' = P_{n+200}$. We may let the hunter know the next 200 feedback points $P_{n+1}, P_{n+2}, \ldots, P_{n+200}$ in the $n + 1$ round beforehand, which will only increase the amount of information the hunter has.

How does the hunter decide after seeing these 200 feedback points? If the hunter moves 200 steps from B_n along $B_n P_{n+200}$, then he will reach B', $B_n B' = 200$. Please note that the hunter has no better strategy! Because, after the hunter moves 200 steps according to any strategy, his position will always be on the left side of B'; if the position is above l, her distance to Y_2 will become larger, otherwise, her distance to Y_1 will become larger. In short, no matter how the hunter chooses her movement method, she cannot guarantee that the distance from the rabbit to her will be less than $y = B'Y_1 = B'Y_2$.

To estimate the value of y^2, let Z be the midpoint of segment Y_1Y_2 and $\varepsilon = ZP'$ (note that $B'P' = d_n$). We have

$$\varepsilon = 200 - A_nZ = 200 - \sqrt{200}^2 - 1 = \frac{1}{200 + \sqrt{200^2 - 1}} > \frac{1}{400}.$$

Particularly, $\varepsilon^2 + 1 = 400\varepsilon$. Then, assuming $d_n < 100$,

$$y^2 = 1 + B'Z^2 = 1 + (d_n - \varepsilon)^2 = d_n^2 - 2\varepsilon d_n + \varepsilon^2 + 1$$
$$= d_n^2 + \varepsilon(400 - 2d_n) > d_n^2 + \frac{1}{2}.$$

Therefore, as long as the positioning device feedbacks such a list of points $P_{n+1}, \ldots, P_{n+200}$, no matter what the hunter does, it is possible to make

$$d_{n+200}^2 > d_n^2 + \frac{1}{2}.$$

So the hunters cannot guarantee that after 10^9 rounds the distance between her and the rabbit is at most 100.

Remark The key to solve this problem is to understand the following:

- From the points given by the positioning device, the hunter can only determine the range of the rabbit's positions. In other words, the same point sequence given by the positioning device can correspond to a variety of different rabbit movements. Therefore, for the hunter, the best strategy is to move in the way that the maximum of her distance to the points in this range is minimized.
- There is a strategy of "coordination" between the rabbit and the positioning device, so that can ensure that the guaranteed minimum distance between the hunter and the rabbit is always increasing at a certain rate.
- A basically correct answer (at least 5 points) should include a computational proof of the following (the interval of 200 steps can be generalized as N):

$$\varepsilon = N - \sqrt{N^2 - 1} > \frac{1}{N + \sqrt{N^2 - 1}} > \frac{1}{2N},$$

and

$$\varepsilon^2 + 1 = 2N\varepsilon.$$

Therefore, as long as $N > d_n$, we get

$$y^2 = d_n^2 + \varepsilon(2N - 2d_n) > d_n^2 + \frac{N - d_n}{N}.$$

For example, take $N = 101$. In this way, we can ensure that the square of the distance between the hunter and the rabbit increases by $\frac{1}{101}$ after every 101 rounds. So it can increase the distance to 100 after $101^2 \times 10^4 = 1.0201 \times 10^8 > 10^9$ rounds. The larger the distance, the smaller the selection range of N.

The original version of the problem provided by Austrian asked if the distance could be controlled within 10^{10} after 10^{100} rounds.

An interesting exercise is to show that "always moving towards the point of the latest feedback from the pointing device" is not necessarily the best strategy for the hunter.

Finally, this very "unconventional" problem became the lowest scoring question in the history of IMO. Two contestants (Russia, Australia) scored 7 points, one (UK) scored 5 points, one scored 4 points, three get 1 point. This is also the first time that the Chinese team has not scored even 1 point on a problem since the 5th problem of the 37th IMO in 1996.

Second Day
(9:00 – 13:00; July 19, 2017)

4 Let R and S be different points on a circle Ω such that RS is not a diameter. Let l be the tangent line to Ω at R. Point T is such that S is the midpoint of the line segment RT. Point J is chosen on the shorter arc $\overset{\frown}{RS}$ of Ω so that the circumcircle Γ of triangle JST intersects l at two distinct points. Let A be the common point of Γ and l that is closer to R. Line AJ meets Ω again at K. Prove that the line KT is tangent to Γ. (Contributed by Luxembourg)

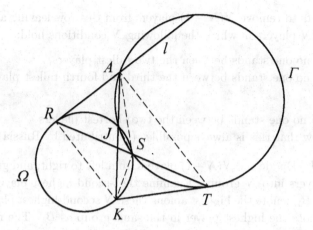

Solution As shown in the figure, since R, K, S, J and S, J, A, T are four concyclic points, respectively, we have

$$\angle KRS = \angle KJS = \angle STA.$$

Because AR is tangent to $\Omega, \angle RKS = \angle TRA$. So $\triangle RKS \sim \triangle TRA$, and then

$$\frac{RK}{RS} = \frac{TR}{TA}.$$

As S is the midpoint of RT, $RS = ST$. So

$$\frac{RK}{TS} = \frac{RT}{TA}.$$

Combing it with $\angle KRT = \angle STA$, we have $\triangle KRT \sim \triangle STA$. Therefore $\angle SAT = \angle STK$, which means line KT is tangent to circle Γ. $\qquad\square$

Remark The scoring standards stipulate as:

- If an inversion method is used, then the step of inverse transformation itself does not give points.
- If the calculation method is used but not completed, then the points can only be given according to the geometric conclusions obtained from the intermediate calculation results written on the paper.

After the test was finished, we noticed that this problem is the same as the third problem of ELMO in 2015.

5 An integer $N > 2$ is given. A collection of $N(N+1)$ soccer players, no two of whom are of the same height, stand in a row. Sir Alex

wants to remove $N(N-1)$ players from this row leaving a new row of $2N$ players in which the following N conditions hold:

(1) no one stands between the two tallest players,
(2) no one stands between the third and fourth tallest players,

...

(N) no one stands between the two shortest players.

Show that this is always possible. (Contributed by Russia)

Solution 1   Divide the $N(N+1)$ players from left to right, and group each $N+1$ players into N groups. Examine the second-highest player in each group, let B_1 denote the highest among these N second-highest players, and let A_1 denote the highest player in the same group as B_1. The remaining $N-1$ players in this group are removed from the team, and the highest players in each of the remaining $N-1$ groups are removed from the team.

In this way, besides A_1 and B_1, the remaining $N(N-1)$ players (their heights are lower than that of B_1) are in the $N-1$ groups. Repeat the same operation for these players, to select A_2 and B_2 from the same group, and then remove $2(N-2)$ players, leaving $(N-1)(N-2)$ players in $N-2$ groups.

After repeating the above operation, we can finally get $2N$ players: A_1, B_1, A_2, $B_2, \ldots, A_N$, B_N. Their height decreases in this order, and each pair of A_i and B_i are in the same group originally divided. There will be no other A_j or B_j between them. So the proposition holds.

Solution 2   Divide all $N(N+1)$ players into N groups according to their heights. The highest $N+1$ players are in the first group, the next $N+1$ players in the second group, ..., the shortest $N+1$ players in the Nth group. Sir Alex *scans* these players from left to right in the row, until the first time he finds there are two players in the same group, assuming it is A_1 and A_2 in the t_1th group, and A_1 is to the left of A_2. Then he lets A_1, A_2 stay, the others on the left of A_2 leave, and everyone else in the t_1th group also leave. The players on the right of A_2 belong to the remaining N-1 groups, and at most one player in each group left. Therefore, there are at least N players in each group. Alex continues to *scan* the remaining players from A_2 to the right until he finds there are two players in the same group again, assuming it is A_3 and A_4 in the t_2th group, and A_3 is on the left of A_4. Then he lets A_3 and A_4 stay, the others between A_2 and A_4 leave, and everyone else in the t_2th group also leave.

This continues. In the end he has $2N$ players $A_1, A_2, \ldots, A_{2N}$, in order from left to right, in the queue. There are exactly two players left in each group, and there are no other players between them in the queue. So the conclusion holds. □

Remark This problem easily led the contestants astray, so the final average score is less than 1 point. It is a typically hard problem. Basically all the correct answers are based on dividing the $N(N+1)$ players into N groups, each group has $N+1$ people, and then using some algorithms to select 2 players from each group.

We should note that each player can be marked with two indicators (p, h): the first one is his position and the second is his height. If we specify: "(a, b) separates (p_1, h_1) and (p_2, h_2)" means "either a is between p_1 and p_2 or b is between h_1 and h_2", then the purpose of Alex is to select $2N$ players $(p_1, h_1), (p_2, h_2), \ldots, (p_{2N}, h_{2N})$ so that no one separates (p_1, h_1) from (p_2, h_2), no one separates (p_3, h_3) from $(p_4, h_4), \ldots$, and no one separates (p_{2N-1}, h_{2N-1}) and (p_{2N}, h_{2N}). From this point of view, the above two methods are dual methods.

In the scoring standard, for a basically complete answer, a maximum of 2 points are deducted for various errors, including

- Minor errors (e.g., it is not proven that there will be no other selected players between the 2 players selected in each group; in the proof using induction / algorithm, fail to confirm that after the induction step the hypothesis condition is still valid; no verification of the induction basis, etc.). All deducted 1 point.
- Major errors (for example, only a correct algorithm is given, but the rationality of the algorithm is not verified). 2 points are deducted.

Steps that are definitely not scored include

- Restate propositions (in the language of geometry, graph theory, matrices, etc.).
- Only make a dual arrangement for the parameters of positions and heights.
- Only verify cases of small $N(= 1, 2, 3, \ldots)$

Considering that this problem is really easy to make mistakes (a dozen coordinators, most of them were former IMO contestants, volunteered to do the problem, only one person scored 7 points, and the rest did not exceed 3 points), so the coordinators gave very loose scoring criteria:

For example, as long as it is clear on the test paper that the $N(N+1)$ players are divided into N groups according to height or position, each

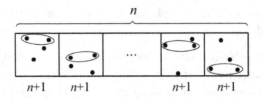

Fig. 5.1

group is $N + 1$, and there is a clear intention to choose 2 from each group players (for example, drawing the following picture on the test paper), 2 points can be obtained.

On this basis, if you can correctly *scan* and select a pair of players as is described in the above, and remove some players correctly (so that the operation can be continued), you can get another 3 points. If you only *scan* and select a pair of players correctly, but the steps to remove players are incorrect, you get 1 more point.

If $2N$ players who meet the requirements are selected from cN^2 players, for $1 < c < 4$, you can get 1 point.

If $2N$ players who meet the requirements are selected from cN^2 players, for $c > 4$ (for example, applying the Erdös-Szekeres theorem), you get 0 points.

We should also note that there are two typical wrong ideas whe n using inductive method to prove this problem: choose two from the leftmost $N+1$ players (so that in the remaining $N^2 - 1$ players there are at most $N - 1$ ones with their heights between the two), and then use the inductive assumption for $N^2 - 1 - (N - 1) = (N - 1)N$ players; or dually, choose two from the highest $N + 1$ players, so that there are only a maximum of $N - 1$ players between them. These two incorrect solutions can only score 1 point.

Another incorrect solution is: First divide the entire players, according to their positons in the row, into N groups $A_1, A_2, \ldots, A_N$, each containing $N + 1$ people; then divide them, according to their heights, into N groups $B_1, B_2, \ldots, B_N$, each containing $N + 1$ people. Use these groups as vertices to construct a bipartite graph with $2N$ vertices, stipulating that A_i and B_j are adjacent if and only if $|A_i \cap B_i| \geq 2$. Then use Hall theorem to find a perfect match, and then take out $2N$ players. 2 points for this incorrect solution.

For interested readers, it is a good exercise to understand exactly what is wrong with these three incorrect solutions.

For $N = 2$ and $N = 3$, $N(N+1)$ is optimal. For example, you can consider 5 players in order of heights (1, 5, 3, 4, 2) or 11 players in order of heights (1, 10, 6, 4, 3, 9, 5, 8, 7, 2, 11).

6 An ordered pair (x, y) of integers is a *primitive point* if the greatest common divisor of x and y is 1. Given a finite set S of primitive points, prove that there exist a positive integer n and integers a_0, $a_1, \ldots, a_n$ such that, for each (x, y) in S, we have:

$$a_0 x^n + a_1 x^{n-1} y + a_2 x^{n-2} y^2 + \ldots + a_{n-1} xy^{n-1} + a_n y^n = 1.$$

(Contributed by the U.S.A.)

Solution 1 First of all, we know that, we only need to find a homogeneous polynomial $f(x, y)$ such that, for any point in $S = \{(x_1, y_1), \ldots, (x_n, y_n)\}$, holds $f(x_i, y_i) = \pm 1 (i = 1, \ldots, n)$ (then $f^2(x, y)$ meets the requirements). If two of these points are co-linear with the origin $(0, 0)$, then they must be symmetrical about the origin, and the absolute value of any homogeneous polynomial at them is the same. Therefore, we may assume that any two points in S are not collinear with the origin.

Consider homogeneous polynomial $l_i(x, y) = y_i x - x_i y$. From the definition of a primitive point, we know $l_i(x_j, y_j) = 0$ if and only if $j = i$. Let

$$g_i(x, y) = \prod_{j \neq i} l_j(x, y).$$

Then $g_i(x, y)$ is an n-1 degree polynomial and has the following two properties:

1) for $j \neq i$, $g_i(x_j, y_j) = 0$;
2) $g_i(x_i, y_i) \neq 0$ (denoted by a_i).

For any integer $N \geq n-1$, there exists a homogeneous polynomial of degree N which also has the above two properties. Just take a polynomial of one degree $I_i(x, y)$ satisfying $I_i(x_i, y_i) = 1$ (because (x_i, y_i) is a primitive point, such I_i always exists), and then consider $I_i(x, y)^{N-(n-1)} g_i(x, y)$.

Let's reduce the problem to the following proposition.

Proposition: For any positive integer a, there is an integer coefficient homogeneous polynomial $f_a(x, y)$ of order no less than 1, so that for any primitive point (x, y), $f_a(x, y) \equiv 1 \pmod{a}$.

To show that the conclusion of the original problem can be obtained from this proposition, we only need to take a as the least common multiple of the aforementioned $a_i (1 \leq i \leq n)$. Take f_a in the proposition and choose one of its powers $(f_a(x,y))^k$ so that its degree is at least $n - 1$, and then subtract a appropriate integer coefficient linear combinazation of $g_i(x,y)$ from this polynomial.

Below we prove the proposition by decomposing a. First, when a is a power of a prime number $(a = p^k)$, then

- if p is an odd prime number, let $f_a(x,y) = (x^{p-1} + y^{p-1})^{\phi(a)}$;
- if $p = 2$, let $f_a(x,y) = (x^2 + xy + y^2)^{\phi(a)}$.

Now suppose a is decomposed into $a = q_1 q_2 \ldots q_k$, where q_i is a power of prime numbers and is coprime with each other. Let f_{q_i} be a polynomial constructed according to the above rules, take one of its appropriate powers F_{q_i}, so that for all i, the degree of F_{q_i} is the same. Note that for any coprime x and y, we have

$$\frac{a}{q_i} F_{q_i}(x, y) \equiv \frac{a}{q_i} (\text{mod } a).$$

According to Bézout's theorem, there exists a linear combination of $\dfrac{a}{q_i}$ with integer coefficients, whose value is exactly 1. Then the same set of coefficients can be used to form a linear combination of of F_{q_i} with integer coefficients, so that for any primitive point (x, y), the value of the polynomial is 1 (mod a). Since the degrees of all F_{q_i} are the same, we get a homogeneous polynomial.

Solution 2 (Based on the original solution 2 to the problem provided by the contributor, and is slightly simplified by Leader of Israeli team Dan Carmon). We use the method of mathematical induction to the number of elements in S. If $|S| = 1$, let $S = \{(x_0, y_0)\}$. According to Bézout's theorem, there are integers a and b such that $ax_0 + by_0 = 1$. Then define integer coefficient polynomial $P(X, Y) = aX + bY$ such that for any $(x, y) \in S$, we have $P(x, y) = 1$.

Assume $|S| = k \geq 2$, and the conclusion holds for $k - 1$. According to Bézout's theorem, for any $(x_0, y_0) \in S$, there exists integers a, b such that $ax_0 + by_0 = 1$. Now define an integer coefficient linear transformation

$$T : \mathbb{R}^2 \to \mathbb{R}^2, T(X, Y) = (aX + bY, -y_0 X + x_0 Y).$$

Then T is also a bijection of $\mathbb{Z}^2$ to $\mathbb{Z}^2$ and maps a primitive point to a primitive point. If there is a homogeneous integer coefficient polynomial $P(X, Y)$, so that for any $(x, y) \in T(S)$, $P(x, y) = 1$, then $P(T(X, Y)) = $

$P(aX + bY, -y_0X + x_0Y)$ is also a homogeneous integer coefficient polynomial and satisfies for any $(x, y) \in S$, $P(T(x, y)) = 1$. So we just need to prove for $W = T(S)$.

Note that $T(x_0, y_0) = (1, 0) \in W$. Let $W' = W \backslash \{(1, 0)\}$. By induction, there is a homogeneous integer coefficient polynomial $F(X, Y)$, so that for any $(x, y) \in W'$, $F(x, y) = 1$. Suppose $W' = \{(x_1, y_1), \ldots, (x_{k-1}, y_{k-1})\}$. Let

$$G(X, Y) = \prod_{i=1}^{k-1}(-x_iY + y_iX),$$

Then we have $G(x_i, y_i) = 0$ for $1 \le i \le k-1$, and $G(1, 0) = y_1y_2\ldots y_{k-1} =: a$. Let

$$F(X, Y) = a_0X^n + a_1X^{n-1}Y + \ldots + a_nY^n.$$

Since $F(x_i, y_i) = a_0x_i^n + y_i(a_ix_i^{n-1} + \ldots + a_ny_i^{n-1}) = 1$, then $(a_0, y_i) = 1$, $1 \le i \le n-1$. Therefore $(a_0, a) = 1$. Take a positive integer d such that $a_0^d \equiv 1 \pmod{a}$ and $d > \deg G$. Let $M = \dfrac{a_0^d - 1}{a} \in Z$, and

$$P(X, Y) = F(X, Y)^d - MX^{d\deg F - \deg G}G(X, Y).$$

Then $P(X, Y)$ is a homogeneous integer coefficient polynomial with degree $d\deg F$. For $1 \le i \le k - 1$,

$$P(x_i, y_i) = F(x_i, y_i)^d - Mx_i^{d\deg F - \deg G}G(x_i, y_i) = 1 - 0 = 1,$$

and

$$P(1, 0) = F(1, 0)^d - MG(1, 0) = a_0^d - \dfrac{a_0^{d-1}}{a} \cdot a = 1.$$

Remark All the solutions to this problem have an algebraic part and a number theory part, and need to establish a connection between them. The scoring standard stipulates that the score of each part should be determined according to the position of the written steps in the corresponding solution.

In Solution 1, if the specific form of $g_i(x, y)$ is written down, get 1 point; if it is shown that for any $d \ge n-1$, there exists a d-th order homogeneous polynomial $h_i(x, y)$, on S that takes the same value as $g_i(x, y)$, get 1 point; if the problem is reduced to the proposition, get 1 point; then the proposition is proved, get 3 points; summarize the conclusion, get 1 point.

In Solution 2, if the specific form of $G(X, Y)$ is written down, get 1 point; if a polynomial of the form $F(X, Y)^k + G(X, Y)H(X, Y)$ is considered, get 1 point. In proving the existence of such k and H:

- If it is proved that for any integer C and $d \geq 0$, such that there is a d–degree polynomial $H(X, Y)$ satisfying $H(x_n, y_n) = C$, 1 point is obtained.
- If the problem is reduced to prove $(G(x_n, y_n), F(x_n, y_n)) = 1$ (e.g. using Euler-Fermat theorem), get 1 point.
- If $(G(x_n, y_n), F(x_n, y_n)) = 1$ is proved, get 2 points; if without the proof but the problem is reduced to consider $(x_n, y_n) = (1, 0)$, get 1 point.
- Summarize the conclusion, get 1 point.

In fact, it is easy to see that the degree of the homogeneous polynomial required cannot be controlled only by the number of elements in the set S, even for the case of two points. Consider $S = \{(1, 0), (a, b)\}$. The required polynomial f satisfies $f(1, 0) = 1$ means that the first coefficient $a_0 = 1$ and

$$1 = f(a, b) \equiv a^{\deg f} \pmod{b}.$$

Therefore, for a properly selected a and b, only $\deg f \geq \phi(b)$ can meet the requirements, but it is obviously unbounded.

The scoring criteria also stipulate:

- if only the case $|S| = 1$ is solved, get 0 point;
- if only the case $|S| = 2$ is solved, get 1 point;
- if is stated in the inductive steps that the last point may be assumed as $(1, 0)$, get 1 point.

The proofs of Ren Qiuyu and Zhang Lu of the Chinese team are different from the official answers. Their thinking is to use the method of mathematical induction based on $|S| = 3$. The induction steps are: Suppose for $|S| = n \geq 3$, the proposition holds. Now for

$$S = \{(x_0, y_0), (x_1, y_1), \ldots, (x_n, y_n)\},$$

first make inductive hypothesis to

$$S_1 = \{(x_1, y_1), \ldots, (x_n, y_n)\},$$

and get that there exists homogeneous binary polynomial $P_1(x, y)$, satisfying $P_1(x_i, y_i) = 1 (i = 1, 2, \ldots, n)$, and let $P_1(x_0, y_0) = a$.

Then make inductive hypothesis to

$$S_2 = \{(x_0, y_0), (x_2, y_2), \ldots, (x_n, y_n)\},$$

and get that there exists homogeneous binary polynomial $P_2(x, y)$, satisfying $P_2(x_i, y_i) = 1(i = 1, 2, \ldots, n)$, and let $P_2(x_0, y_0) = b$.

If one of a and b is equal to 1, then the problem is solved. Otherwise, by the basis of induction (the case $n = 3$), there is a homogeneous binary polynomial $Q_0(x, y)$ satisfying

$$Q_0(a, 1) = Q_0(1, b) = Q_0(1, 1) = 1.$$

Then

$$P(x, y) = Q_0(P_1(x, y), P_2(x, y))$$

satisfies the requirement.

Confirmed by the coordinators, this step is reduced to the process of $n = 3$ and 3 points are given.

Let's take a look at Ren Qiuyu's proof of $n = 3$ (unfortunately, he didn't write it completely because of insufficient time in the test room).

Lemma *For any $a, b \in Z \backslash \{1\}$, there is an integer coefficient homogeneous polynomial $h(x, y)$ such that $h(a, 1) = h(1, b) = h(1, 1) = 1$.*

Proof If $|ab| = 1$, then $|a| = |b| = 1$. Just let

$$h(x, y) = 2x^2 - y^2.$$

Suppose for $|ab| \neq 1$, we have

$$h(x, y) = k_0 x^n + k_1 x^{n-1} y + \ldots + k_n y^n$$

satisfying

$$k_0 a^n + k_1 a^{n-1} + \ldots + k_n = 1,$$
$$k_0 + k_1 b + \ldots + k_n b^n = 1,$$
$$k_0 + k_1 + \ldots + k_n = 1.$$

There are 3 equations and $n + 1$ unknowns. From the last equation,

$$k_0 = 1 - k_1 - \ldots - k_n.$$

Substituting it into the other two equations, we have, respectively,

$$\sum_{i=1}^{n} k_i(a^{n-i} - a^n) = 1 - a^n$$

$$\Leftrightarrow \sum_{i=1}^{n} k_i a^{n-i}(1 + a + \ldots + a^{i-1}) = 1 + a + \ldots + a^{n-1}. \tag{1}$$

$$\sum_{i=1}^{n} k_i(b^i - 1) = 0 \Leftrightarrow \sum_{i=1}^{n} k_i(1 + b + \ldots + b^{i-1}) = 0$$

$$\Rightarrow k_1 = -\sum_{i=2}^{n} k_i(1 + b + \ldots + b^{i-1}). \tag{2}$$

Then

$$\sum_{i=1}^{n} k_i(a^{n-i} + \ldots + a^{n-1} - a^{n-1} - a^{n-1}b - \ldots - a^{n-1}b^{i-1})$$

$$= 1 + a + \ldots + a^{n-1},$$

and that is

$$(1 - ab) \sum_{i=2}^{n} k_i[a^{n-2} + a^{n-3}(1 + ab) + \ldots + a^{n-i}(1 + ab + \ldots + a^{i-2}b^{i-2})]$$

$$= 1 + a + \ldots + a^{n-1}. \tag{3}$$

We note that the coefficient of k_2 is $M = (1 - ab)a^{n-2}$, and that of k_n is

$$N = (1 - ab)[a^{n-2} + a^{n-3}(1 + ab) + \ldots + (1 + ab + \ldots + a^{n-2}b^{n-2})].$$

Since $(a, N) = (a, 1) = 1$, we get $(M, N) = 1 - ab$. Therefore, the indefinite linear equation ③ must have an integer solution, as long as $(1 - ab) \,|\, 1 + a + \ldots + a^{n-1}$, which is equivalent to $(1 - ab)(1 - a) \,|\, a^{n-1}$, since $a \neq 1$. As $(a, (1 - ab)(1 - a)) = 1$, we can take $n = \varphi(|(1 - ab)(1 - a)|)$. The proof of the lemma is completed.

For three primitive points at general positions, we know that there is a single-mode transformation with integer coefficients, mapping one of these points to $(1, 1)$. Then do a translation transformation $(x, y) \to (x - 1, y - 1)$, and then do a single mode integer coefficient linear transformation to change the other two points to the coordinate axis. Finally do a translation transformation $(x, y) \to (x+1, y+1)$. Then it is reduced to the lemma case.

With a similar discussion, we can prove that the proposition holds when $|S| = 2$.

International Mathematical Olympiad

2018 (Cluj-Napoca, Romania)

The 59th International Mathematical Olympiad (IMO) was held in Cluj-Napoca, Romania from July 3 to 14, 2018. 594 students from 107 countries and regions participated in the competition. The Chinese team won the third place in the group with a total of 199 points. The 6 players won 4 gold medals and 2 silver medal.

The members of the Chinese team are as follows:

Team Leader: Qu Zhenhua, East China Normal University

Deputy Leader: He Yijie, East China Normal University

Observers: Xiong Bin, East China Normal University

Wang Bin, Chinese Academy of Sciences Academy of Mathematics and Systems Science

Zhang qi, Youth Science and Technology Center of China Association for Science and Technology

Team Contestants:

Chen Yinyi, Changsha City Yali Middle School in Hunan Province; 37 points, Gold Medal

Ouyang Zexuan, Wenzhou Middle School in Zhejiang Province; 36 points, Gold Medal

Li Yixiao, Tianyi High School in Jiangsu Province; 35 points, Gold Medal

Wang Zeyu, Middle School Affiliated to Shanxi Northwestern Polytechnical University; 31 points, Gold Medal

Yao Rui, The First Affiliated High School of Central China Normal University, Hubei Province; 30 points, Silver Medal

Ye Qi, Yueqing City Yuecheng Boarding High School, Zhejiang Province; 30 points, Silver Medal

The top ten teams and their total scores are as follows:

1. United States;   212 points
2. Russia;   201 points
3. China;   199 points
4. Ukraine;   186 points
5. Thailand;   183 points
6. Taiwan, China   179 points
7. South Korea;   177 points
8. Singapore;   175 points
9. Poland;   174 points
10. Indonesia;   171 points

The gold medal line is 31 points, the silver medal line is 25 points, and the bronze medal line is 16 points.

The Six contestants of the Chinese National Team received training at East China Normal University, Shanghai, from June 15 to 25. Qu Zhenhua, Xiong Bin, Leng Gangsong, Yu Hongbing, He Yijie, Wang Bin, Zhang Sihui, and Fu Yunhao provided tutoring to them. Lin Tianqi and Yang Xinyuan served as the head teachers. Teacher Jia Zhi, secretary of the Party Committee of the School of Mathematical Sciences, and Teacher Zhang Hongyan from the office did a lot of work for the national team training. From July 2nd to 7th, the national team conducted a pre-departure training session at the School of Mathematical Sciences, Peking University. Secretary-General of the Chinese Mathematical Society, Professor Chen Dayue, Dean of the School of Mathematical Sciences, Peking University, Professor Liu Bin, Former Associate Dean of Teaching, and Professor Li Ruo, Associate Dean of Teaching, Deputy Dean of the Mathematical Olympiad Committee of the Chinese Mathematical Society Professor Wu Jianping, the director, had a discussion with the students. Professor Qiu Zonghu received the students and gave them pre-match counseling. Teacher Sun Zhaojun, deputy secretary of the Party Committee of the School of Mathematical Sciences of Peking University, also did a lot of work for the national team.

First Day
(9:30 – 14:00; July 9, 2018)

1 Let Γ be the circumcircle of acute-angled triangle ABC. Points D and E lie on segments AB and AC, respectively, such that $AD = AE$. The perpendicular bisectors of BD and CE intersect the minor arcs AB and AC of Γ at points F and G, respectively. Prove that the lines DE and FG are parallel (or are the same line).

Solution   As shown in Fig. 1.1, P, Q, R are the midpoints of minor arcs $\overset{\frown}{BC}$, $\overset{\frown}{CA}$, $\overset{\frown}{AB}$ on Γ, respectively, M, N are the midpoints of AB, AC, respectively, and O is the centre of circle Γ. Then O, M, R are collinear, and O, N, Q are collinear.

Since $AD = AE$, AP bisects $\angle BAC$, and then $AP \perp DE$. Furthermore,

$$\overset{\frown}{AQ} + \overset{\frown}{PBR} = \frac{\angle B}{2} + \frac{\angle A + \angle C}{2} = 90°.$$

Hence, $QR \perp AP$, and $DE \parallel RQ$. Then we only need prove that $FG \parallel RQ$, and that is equivalent to prove $\overset{\frown}{FR} = \overset{\frown}{GQ}$. Let F', G' be the midpoints of segments BM, CN, respectively. Then $FF' \perp AB, GG' \perp AC$, so $FF' \parallel OR$, $GG' \parallel OQ$. We have

$$F'M = BM - BF' = \frac{1}{2}(AB - BD) = \frac{AD}{2} = \frac{AE}{2} = G'N.$$

Then the distance between lines FF' and OR is equal to that between GG' and OQ. Therefore, $\overset{\frown}{FR} = \overset{\frown}{GQ}$. The proof is completed. $\square$

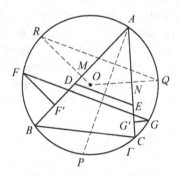

Fig. 1.1

2 Find all integers $n \geq 3$ for which there exist real numbers a_1, $a_2, \ldots, a_{n+2}$, such that $a_{n+1} = a_1$ and $a_{n+2} = a_2$, and

$$a_i a_{i+1} + 1 = a_{i+2}$$

for $i = 1, 2, \ldots, n$.

Solution 1   We will prove that n is a multiple of 3.

On the one hand, let $n = 3k$, $a_{3i-2} = a_{3i-1} = -1$, $a_{3i} = 2$, for $i = 1, 2, \ldots, k$, and $a_{n+1} = a_{n+2} = -1$. Then it is easy to check that $a_i a_{i+1} + 1 = a_{i+2}$, $i = 1, 2, \ldots, n$, meeting the required condition.

On the other hand, suppose there are $a_1, a_2, \ldots, a_{n+2}$ satisfying the required condion. We extend them to an infinite sequence of period n, without end in both directions. Then for any integer i, $a_i a_{i+1} + 1 = a_{i+2}$. We will prove the following statements:

(1) There is no integer i, such that $a_i > 0$, $a_{i+1} > 0$.

Otherwise, assume there is an integer i, such that $a_i > 0$, $a_{i+1} > 0$. Then $a_{i+2} = a_i a_{i+1} + 1 > 1$. It is easy to show by induction that $a_m > 1$ for every $m \geq i + 2$. Then for $m \geq i + 2$,

$$a_{m+2} = a_m a_{m+1} + 1 > a_m a_{m+1} > a_{m+1}.$$

Hence, $a_{m+1} < a_{m+2} < a_{m+3} < \cdots < a_{m+n+1}$, a contradiction to $a_{m+n+1} = a_m$.

(2) There is no integer i, such that $a_i = 0$.

Otherwise, assume there is i, such that $a_i = 0$. Then $a_{i+1} = a_{i-1} a_i + 1 = 1$, $a_{i+2} = a_i a_{i+1} + 1 = 1$, a contradiction to (1).

(3) There is no i, such that $a_i < 0$, $a_{i+1} < 0$, $a_{i+2} < 0$.

Assume there is i, such $a_i < 0$, $a_{i+1} < 0$, $a_{i+2} < 0$. Then $a_{i+2} = 1 + a_i a_{i+1} > 0$, a contradiction.

(4) There is no i, such that $a_i > 0$, $a_{i+1} < 0$, $a_{i+2} > 0$.

Assume there is i, such that $a_i > 0$, $a_{i+1} < 0$, $a_{i+2} > 0$. Then from (1) we get $a_{i-1} < 0$, $a_{i+3} < 0$. Since $0 < a_{i+2} = 1 + a_i a_{i+1} < 1$ and $|a_{i+1} a_{i+2}| = |a_{i+3} - 1| > 1$, then $|a_{i+1}| > 1$ and $a_{i+1} < -1$. From $|a_i a_{i+1}| = |a_{i+2} - 1| < 1$, we get $|a_i| < 1$, and then $0 < a_i < 1$. Furthermore, from $a_{i-2} a_{i-1} = a_{i-1} < 0$, we get $a_{i-2} > 0$. From $a_{i+1} < 0$ and $0 < a_i < 1$, we have

$$a_{i-1} = \frac{a_{i+1} - 1}{a_i} < a_{i+1} - 1 < a_{i+1}.$$

Hence, from $a_i > 0$, $a_{i+1} < 0$, $a_{i+2} > 0$, we get $a_{i-1} < 0$, $a_{i-2} > 0$, and $a_{i-1} < a_{i+1}$. By induction it is easy to know that for any $k \geq 0$, we have

$a_{i-2k} > 0$, $a_{i-2k+1} < 0$, and

$$a_{i+1} > a_{i-1} > a_{i-3} > a_{i-5} > \ldots.$$

In particular, $a_{i+1} > a_{i+1-2n}$, and it is a contradication to the definition that $\{a_m\}_{m\in\mathbb{Z}}$ is an infinite sequence of period n.

From (1), (2), (3), (4) we know that in any consecutive terms a_i, a_{i+1}, a_{i+2}, exactly one of them is positive and the two are negative. Hence, the plus or minus sign in the sequence has a period of 3. In particular, the plus sign does. Therefore, if $a_i > 0$, then $a_{i+n} > 0$, which means $3 \mid n$.

Solution 2 We will give another proof of $3 \mid n$. Extend $a_1, a_2, \ldots, a_n$ to an infinite sequence of period n, without end in both directions. Then for any integer i, we have

$$a_i a_{i+1} a_{i+2} = (a_{i+2} - 1)a_{i+2} = a_{i+2}^2 - a_{i+2},$$
$$a_i a_{i+1} a_{i+2} = a_i(a_{i+3} - 1) = a_i a_{i+3} - a_i.$$

Hence,

$$a_{i+2}^2 - a_{i+2} = a_i a_{i+3} - a_i.$$

Suming the above formula for $i = 1, 2, \ldots, n$, we have

$$\sum_{i=1}^{n} a_i^2 = \sum_{i=1}^{n} a_i a_{i+3}.$$

It is equivalent to

$$\sum_{i=1}^{n} (a_i - a_{i+3})^2 = 0.$$

Then for any i, $a_i = a_{i+3}$, which means $\{a_i\}$ has a period of 3. If $3 \nmid n$, then $\{a_i\}$ has a period of 1, i.e., a constant sequence. At this time, however, equation $x^2 + 1 = x$ has no real solution. So there is no constant sequence that satisfies the required condition of the problem. Therefore, $3 \mid n$. □

Remark In actual competitions, most students have solved the problem in a similar way to Solution 1, i.e., to prove that the plus-minus signs in the sequence have a period of 3. Solution 2 uses a method of overall inequality. Although it is short and simple, it is not easy to think of.

③ An anti-Pascal triangle is an equilateral triangular array of numbers such that, except for the numbers in the bottom row, each number is the absolute value of the difference of the two numbers immediately below it. For example, the following array is an anti-Pascal triangle with four rows which contains every integer from 1 to 10.

$$4$$

$$2 \quad 6$$

$$5 \quad 7 \quad 1$$

$$8 \quad 3 \quad 10 \quad 9$$

Does there exist an anti-Pascal triangle with 2018 rows which contains every integer from 1 to $1 + 2 + \cdots + 2018$?

Solution 1 We will prove by contradication that such a kind of anti-Pascal triangle doesn't exist.

Assuming there is an anti-Pascal triangle with the required property, let $N = 1 + 2 + \ldots + 2018$, and record the j-th number from left to right in the i-th row as $a_{i,j}$. Let $a_1 = a_{1,1}$, and record the larger of the two numbers below a_1 as a_2, and the smaller as b_2; record the larger of the two numbers below a_2 as a_3, and the smaller as b_3; and so on; record the larger of the two numbers below a_i as a_{i+1}, and the smaller as b_{i+1}, $i = 1, 2, \ldots, 2017$.

According to the definition of anti-Pascal triangle, we have $a_i = a_{i+1} - b_{i+1}$, $i = 1, 2, \ldots, 2017$. Hence,

$$a_{2018} = a_1 + b_2 + b_3 + \cdots + b_{2018}.$$

As $a_1, b_2, b_3, \ldots, b_{2018}$ are 2018 mutually different positive numbers,

$$a_1 + b_2 + b_3 + \cdots + b_{2018} \geq 1 + 2 + 3 + \cdots + 2018 = N.$$

But $a_{2018} \leq N$, then $a_{2018} = N$ and

$$\{a_1, b_2, b_3, \ldots, b_{2018}\} = \{1, 2, 3, \ldots, 2018\}.$$

Let $a_{2018} = a_{2018,j}$. By symmetry, we may assume $j \leq 1009$.

$$a_{2018}, b_{2018} \in \{a_{2018,1}, a_{2018,2}, \ldots, a_{2018,1010}\}.$$

Therefore, $a_1, a_2, \ldots, a_{2018}, b_2, b_3, \ldots, b_{2018}$ are all in the following set S:

$$S = \{a_{i,j} \mid j \leq 1010\}.$$

Consider the remaining numbers in the anti-Pascal triangle

$$T = \{a_{i,j} \mid 1011 \leq i \leq 2018, \quad 1011 \leq j \leq i\}.$$

Let $c_{2011} = a_{1011,1011}$. For $i = 1011, 1012, \ldots, 2017$, record the larger of the two numbers below c_i as c_{i+1}, and the smaller as d_{i+1}. Since $c_i = c_{i+1} - d_{i+1}$,

$i = 1011, 1012, \ldots, 2017$, we have

$$c_{2018} = c_{1011} + d_{1012} + d_{1013} + \cdots + d_{2018}.$$

As $c_{1011}, d_{1012}, d_{1013}, \ldots, d_{2018}$ are all in T and not in S, they are all greater than 2018 and mutually different. Therefore,

$$c_{1011} + d_{1012} + d_{1013} + \cdots + d_{2018} \geq 2019 + 2020 + \cdots + 3026$$

$$= 2542680 > N = 2037171.$$

That contradicts to $c_{2018} \leq N$. The proof is completed.

Solution 2 Also by contradiction, assume there is an anti-Pascal triangle with the required property. Let $N = \dfrac{1}{2} \times 2018 \times 2019$. We call $1, 2, \ldots, 2018$ as small numbers, and $N, N-1, \ldots, N-2018$ as big numbers. Note that, for each big number not in the last row, the two numbers below it are exactly one big and one small. We are going to prove the following three statements.

(1) There is exactly one small number in each row. Prove: Let a_i, b_i be the largest and smallest numbers in the i row, respectively. For $1 \leq i \leq 2017$, let the two numbers below a_i are b, c, with $b > c$. Then

$$a_i = b - c \leq a_{i+1} - b_{i+1}.$$

Summing it for $i = 1, 2, \ldots, 2017$, we have

$$a_1 \leq a_{2018} - \sum_{i=2}^{2018} b_i.$$

Since $a_1, b_2, b_3, \ldots, b_{2018}$ are mutually different positive integers,

$$a_{2018} \geq a_1 + \sum_{i=2}^{2018} b_i \geq \sum_{i=1}^{2018} i = N.$$

But $a_{2018} \leq N$, then $a_{2018} = N$ and

$$\{a_1, b_2, b_3, \ldots, b_{2018}\} = \{1, 2, \ldots, 2018\}.$$

(2) Big numbers are all in the last 80 rows. Prove: for $i \leq 1938$, we have

$$a_i \leq a_{2018} - \sum_{j=i+1}^{2018} b_j \leq N - \sum_{j=1}^{2018-i} j \leq N - \sum_{j=1}^{80} j = N - 3240.$$

Hence, a_i is not a big number, and there are not any big numbers in the first 1938 rows.

(3) Statements (1) and (2) lead to a contradiction. Prove: If there are two large numbers adjacent to each other in the last row, then the number above them must be a small one. Since there is only one small number in the penultimate row, there are at most two big numbers adjcacent to each other in the last row. Then the number of big numbers in the last row is not greater than $2 + \left\lceil \dfrac{2016}{2} \right\rceil = 1010$. For each big number not in the last row, there must be a small one in the two numbers below it. Since all the big numbers are in the last 80 rows and there is at least one small number in each row, the number of big numbers that are not in the last row is not greater than $79 \times 2 = 158$. Therefore, the total number of big numbers is not greater than $1010 + 158 = 1168$, contradicting to the fact that there are totally 2019 big numbers.

     The proof is completed.   □

Second Day
(9:30 – 14:00; July 10, 2018)

4   A site is any point (x, y) in the plane such that x and y are both positive integers less than or equal to 20. Initially, each of the 400 sites is unoccupied. Amy and Ben take turns placing stones with Amy going first. On her turn, Amy places a new red stone on an unoccupied site such that the distance between any two sites occupied by red stones is not equal to $\sqrt{5}$. On his turn, Ben places a new blue stone on any unoccupied site. (A site occupied by a blue stone is allowed to be at any distance from any other occupied site.) They stop as soon as a player cannot place a stone. Find the greatest K such that Amy can ensure that she places at least K red stones, no matter how Ben places his blue stones.

Solution   We will prove that $K = 100$.

     On the one hand, Amy has a strategy to ensure that she can place at least 100 red stones. We divid all the sites into two types: odd sites where $2 \nmid x + y$ and even sites that are not odd. As the distance between any two odd sites is not equal to $\sqrt{5}$ and there are totally 200 odd sities. Amy can place a red stone on an unoccupied odd site on each of her first 100 turns. Therefore, $K \geq 100$.

     On the other hand, Ben has a strategy to ensure that Amy cannot place 101 red stones. We consider a 4×4 matrix of sites, where each site is marked with a letter as shown in the following.

$$\begin{array}{cccc} A & B & C & D \\ C & D & A & B \\ B & A & D & C \\ D & C & B & A \end{array}$$

Then divid the sites in the matrix into four groups: each group consists of the four sites marked with the same letter. In each group, linking any two sites that have a distance of $\sqrt{5}$ between them, then we get a parallelogram. In this way, we have 100 parallelograms composed by the 400 sites. Each time after Amy chooses a site P in a pallelogram to places a red stone, Ben just place a blue one on the site opposite to P in the pallelogram. Then Amy cannot place another red stone in this pallelogram. Therefore, Amy can place at most one red stone in every pallelogram, and then she cannot place 101 red stone in whole.

In summary, $K = 100$. The proof is completed. □

5 Let $a_1, a_2, \ldots$ be an infinite sequence of positive integers. Suppose that there is an integer $N > 1$ such that, for each $n \geq N$, the number

$$\frac{a_1}{a_2} + \frac{a_2}{a_3} + \cdots + \frac{a_{n-1}}{a_n} + \frac{a_n}{a_1}$$

is an integer. Prove that there is a positive integer M such that $a_m = a_{m+1}$ for all $m \geq M$.

Solution By the given condition, we know that, for integer $n \geq N$,

$$\left(\frac{a_1}{a_2} + \frac{a_2}{a_3} + \cdots + \frac{a_n}{a_{n+1}} + \frac{a_{n+1}}{a_1} \right) - \left(\frac{a_1}{a_2} + \frac{a_2}{a_3} + \cdots + \frac{a_{n-1}}{a_n} + \frac{a_n}{a_1} \right)$$

$$= \frac{a_n}{a_{n+1}} + \frac{a_{n+1}}{a_1} - \frac{a_n}{a_1}$$

is an integer. Then for $n \geq N$, $\dfrac{a_1 a_n}{a_{n+1}} + a_{n+1} - a_n$ is an integer, i.e., $a_{n+1} \mid a_1 a_n$. By induction we get, for $n > N$, $a_n \mid a_1^{n-N} a_N$. Let P be the set of all the prime factors of $a_1 a_N$. Then P is a finite set. For $n > N$, the prime factors of a_n are all in P, because $a_n \mid a_1 a_N^{n-N}$.

In order to prove that $\{a_n\}_{n \geq 1}$ is a constant in the end, we only need to prove that, for every prime $p \in P$, $\{v_p(a_n)\}_{n \geq 1}$ is a constant in the end.

Let $p \in P$, $n \geq N$. Then one and only one of the following two statements is true:

(1) $v_p(a_{n+1}) \leq v_p(a_n)$;

(2) $v_p(a_{n+1}) > v_p(a_n)$, and $v_p(a_{n+1}) = v_p(a_1)$.

As a fact of matter, if $v_p(a_{n+1}) > v_p(a_n)$, then $v_p\left(\dfrac{a_n}{a_{n+1}}\right) < 0$ and $v_p\left(\dfrac{a_n}{a_1}\right) < v_p\left(\dfrac{a_{n+1}}{a_1}\right)$. We know that

$$v_p\left(\frac{a_n}{a_{n+1}} + \frac{a_{n+1}}{a_1} - \frac{a_n}{a_1}\right) \geq 0.$$

Combing it with $v_p\left(\dfrac{a_n}{a_{n+1}}\right) < 0$, we get that the minimum value in $v_p\left(\dfrac{a_n}{a_{n+1}}\right)$, $v_p\left(\dfrac{a_n}{a_1}\right)$ and $v_p\left(\dfrac{a_{n+1}}{a_1}\right)$ appears at least twice, so

$$v_p\left(\frac{a_n}{a_{n+1}}\right) = v_p\left(\frac{a_n}{a_1}\right).$$

That is $v_p(a_{n+1}) = v_p(a_1)$. Next we will prove

(3) If $v_p(a_n) = v_p(a_1)$, where $n > N$, $p \in P$, then $v_p(a_{n+1}) = v_p(a_1)$.

Assuming $v_p(a_{n+1}) \neq v_p(a_1)$, from (1) and (2) we get $v_p(a_{n+1}) < v_p(a_n) = v_p(a_1)$. Hence,

$$v_p\left(\frac{a_n}{a_{n+1}}\right) > 0, \quad v_p\left(\frac{a_n}{a_1}\right) = 0, \quad v_p\left(\frac{a_{n+1}}{a_1}\right) < 0.$$

Therefore,

$$v_p\left(\frac{a_n}{a_{n+1}} + \frac{a_{n+1}}{a_1} - \frac{a_n}{a_1}\right) = v_p\left(\frac{a_{n+1}}{a_1}\right) < 0,$$

contradicting to the fact that $\dfrac{a_n}{a_{n+1}} + \dfrac{a_{n+1}}{a_1} - \dfrac{a_n}{a_1}$ is an integer.

For any $p \in P$, we have the following two cases:

Case 1: $v_p(a_{n+1}) \leq v_p(a_n)$ for any $n > N$. Then $\{v_p(a_n)\}_{n \geq N}$ is a monotonic non-increasing sequence of non-negative integers, so it will be a constant in the end.

Case 2: There exists $n_0 \geq N$ such that $v_p(a_{n_0+1}) > v_p(a_{n_0})$. From (2) we have $v_p(a_{n_0+1}) = v_p(a_1)$. Then from (3) and by induction we know that $v_p(a_n) = v_p(a_1)$, for any $n \geq n_0$. Therefore, $\{v_p(a_n)\}$ is a constant in the end.

6   A convex quadrilateral $ABCD$ satisfies $AB \cdot CD = BC \cdot DA$. Point X lies inside $ABCD$ so that

$$\angle XAB = \angle XCD \text{ and } \angle XBC = \angle XDA.$$

Prove that $\angle BXA + \angle DXC = 180°$.

Solution 1   We only need to prove that

$$\frac{XB}{XD} = \frac{AB}{CD},$$

①

and

$$\frac{XA}{XC} = \frac{DA}{BC}.$$

②

Because from Equation ① and the sine theorem, we have

$$\frac{\sin \angle AXB}{\sin \angle XAB} = \frac{AB}{XB} = \frac{CD}{XD} = \frac{\sin \angle CXD}{\sin \angle XCD}.$$

Then by the given condition $\angle XAB = \angle XCD$, we get

$$\sin \angle AXB = \sin \angle CXD.$$

In the same way, from Equation ② we obtain

$$\sin \angle DXA = \sin \angle BXC.$$

If either $\angle AXB + \angle CXD = 180$ or $\angle DXA + \angle BXC = 180°$, the problem is solved. Otherwise,

$$\angle AXB = \angle CXD \text{ and } \angle DXA = \angle BXC,$$

which means X is the intersection of lines AC and BD. By the given conditions, we know that $ABCD$ is a pallelogram as well as a rhombus. Then $AC \perp BD$, and the problem is also solved.

In order to prove Equations ① and ②, we take X as the center with radius 1 to do the inversion transformation, as shown in Figure 1, where A', B', C', D' represent the inverted images of A, B, C, D, respectively.

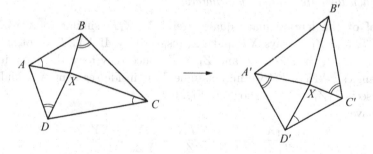

Fig. 6.1

Since $XA \cdot XA' = XB \cdot XB' = XC \cdot XC' = XD \cdot XD'$, then

$$\triangle XAB \backsim \triangle XB'A' \text{ and } \triangle XBC \backsim \triangle XC'B';$$

$$\angle XB'A' = \angle XAB = \angle XCD \text{ and } \angle XCB = \angle XB'C'.$$

Hence,

$$\angle BCD = \angle BCX + \angle XCD = \angle XB'C' + \angle A'B'X = \angle A'B'C'.$$

In a similar way,

$$\angle CDA = \angle B'C'D', \angle DAB = \angle C'D'A', \text{ and } \angle ABC = \angle D'A'B'.$$

Therefore, the corresponding internal angles of quadrilaterals $ABCD$ and $D'A'B'C'$ are equal. By the properties of similar triangles, we get

$$\frac{A'B'}{AB} = \frac{XB'}{XA} = \frac{1}{XA \cdot XB}.$$

Hence, $A'B' = \dfrac{AB}{XA \cdot XB}$. In a similar way, we get the formulas for $B'C'$, $C'D'$, $D'A'$. Then we have

$$A'B' \cdot C'D' = \frac{AB}{XA \cdot XB} \cdot \frac{CD}{XC \cdot XD} = \frac{BC}{XB \cdot XC} \cdot \frac{DA}{XD \cdot XA}$$
$$= B'C' \cdot D'A'.$$

Thererfore, quadrilaterals $ABCD$ and $D'A'B'C'$ have the same internal angles and the same property of equal products of opposite sides.

Next, we need the following lemma to finish the proof.

Lemma *Let $XYZT$ and $X'Y'Z'T'$ be two convex quadrilaterals, their corresponding internal angles be equal, and satisfy*

$$XY \cdot ZT = YZ \cdot TX, X'Y' \cdot Z'T' = Y'Z' \cdot T'X'.$$

Then they are similar quadrilaterals.

Proof of the lemma make quadrilateral XYZ_1T_1 similar to $X'Y'Z'T'$, with T_1 and Z_1 on rays XT and YZ, respectively. If $XYZT$ is not similar to $X'Y'Z'T'$, then $T_1 \neq T$ and $Z_1 \neq Z$. Since they have the same internal angles, $T_1Z_1 \parallel TZ$ we may assume T_1 is inside segment XT, and XZ intersects T_1Z_1 at U, as shown in Fig. 6.2.
We have

$$\frac{T_1X}{T_1Z_1} < \frac{T_1X}{T_1U} = \frac{TX}{ZT} = \frac{XY}{YZ} < \frac{XY}{YZ_1}.$$

Hence, $T_1X \cdot YZ_1 < T_1Z_1 \cdot XY$, acontradiction. The proof is completed.

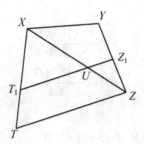

Fig. 6.2

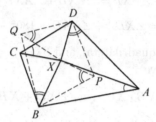

Fig. 6.3

Returning to the original problem, we have proved that quadrilaterals $ABCD$ and $D'A'B'C'$ are similar. Then

$$\frac{BC}{AB} = \frac{A'B'}{D'A'} = \frac{AB}{XA \cdot XB} \cdot \frac{XD \cdot XA}{DA} = \frac{AB}{AD} \cdot \frac{XD}{XB},$$

Therefore,

$$\frac{XB}{XD} = \frac{AB^2}{BC \cdot AD} = \frac{AB^2}{AB \cdot CD} = \frac{AB}{CD}.$$

Then we complete the proof of Equation ①. In the same way, we are able to prove Equation ②. The whole proof is completed.

Solution 2 By contradiction, assume the conclusion in the problem is not true. Without loss of generality, suppose $\angle AXD + \angle BXC < 180°$. As shown in Fig. 6.3, make points P and Q such that $\triangle BXP \backsim \triangle AXD$ and $\triangle DXQ \backsim \triangle CXB$. Then

$$\frac{BX}{AX} = \frac{XP}{XD} = \frac{BP}{AD}.$$

Hence,

$$XP = \frac{XB \cdot XD}{XA},$$ (3)

$$BP = \frac{XB \cdot AD}{XA}.$$ (4)

We may assume

$$AB \cdot CD = BC \cdot AD = 1.$$ (5)

Since

$$\angle CXB + \angle BXP = \angle CXB + \angle AXD < 180°,$$

$$\angle XBC + \angle XBP = \angle XDA + \angle XAD = 180° - \angle AXD < 180°,$$

then $CBPX$ is a convex quadrilateral, and $S_{\triangle BCP} < S_{\triangle BCX} + S_{\triangle BPX}$. By the sine theorem, we have

$$BC \cdot BP \sin \angle AXD < XB \cdot XC \cdot \sin \angle BXC + XB \cdot XP \cdot \sin \angle AXD.$$ (6)

Substitute ③, ④ into ⑥, and use ⑤ to get

$$\sin \angle AXD < XA \cdot XC \cdot \sin \angle BXC + XB \cdot XD \cdot \sin \angle AXD.$$ (7)

In the same way, $ADQX$ is also a convex quadrilateral, $S_{\triangle ADQ} < S_{\triangle ADX} + S_{\triangle DQX}$, and

$$\sin \angle BXC < XA \cdot XC \cdot \sin \angle AXD + XB \cdot XD \cdot \sin \angle BXC.$$ (8)

Adding ⑦ and ⑧, and eleminating $\sin \angle AXD + \sin \angle BXC > 0$, we get

$$XA \cdot XC + XB \cdot XD > 1.$$ (9)

Since $\angle PXD = \angle BXA$ and $\dfrac{XP}{XD} = \dfrac{BX}{AX}$, we have $\triangle BXA \backsim \triangle PXD$. Hence,

$$\angle XDP = \angle XAB.$$

Furthermore, since

$$\angle CXD + \angle DXP > 180°,$$

$CDPX$ is a concave quadrilateral, and then $S_{\triangle CDP} > S_{\triangle CDX} + S_{\triangle DPX}$. By the since theorem, we get

$$\sin \angle CXD > XA \cdot XC \cdot \sin \angle CXD + XB \cdot XD \cdot \sin \angle AXB.$$ (10)

In the same way, $\triangle DXC \backsim \triangle QXB$, $ABQX$ is a concave quandrilateral, and $S_{\triangle ABQ} > S_{\triangle ABX} + S_{\triangle BQX}$. By the sine theorem, we get

$$\sin \angle AXB > XA \cdot XC \cdot \sin \angle AXB + XB \cdot XD \cdot \sin \angle CXD. \quad (11)$$

Adding (10) and (11), we have

$$XA \cdot XC + XB \cdot XD < 1. \quad (12)$$

It is a contradiction to (9). The proof is completed. $\qquad \square$

Printed in the United States
by Baker & Taylor Publisher Services

Printed in the United States
by Baker & Taylor Publisher Services